HARPIA
PUBLISHING+
AF269419

Unmanned Combat Aerial Vehicles

Current Types, Ordnance and Operations

Dan Gettinger

Strategic Handbook Series | Volume 1

Unmanned Combat Aerial Vehicles

Current Types, Ordnance and Operations

Dan Gettinger

HARPIA PUBLISHING +

Consulting and inspiration by Kerstin Berger

Inside front cover: a US Air Force MQ-9 Reaper armed with AGM-114 Hellfire missiles (USAF/Haley Stevens)

Editorial by Thomas Newdick

Layout by Norbert Novak

Map by James Lawrence, Gingercake Creative

Printed at Finidr, Czech Republic

Harpia Publishing is a member of

ISBN 978-1-950394-05-0

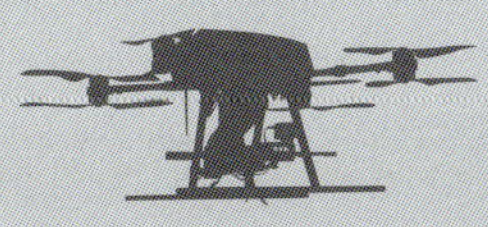

Contents

Introduction

From the earliest days of aviation, militaries have sought to use uncrewed aircraft to their advantage. During the 1849 siege of Venice in the Italian War of Independence, Austrian forces remotely triggered the release of bombs from unmanned balloons floating over the city.[1] In 1910, an inventor in New York envisioned launching small remotely controlled armed drones from a Zeppelin airship, excitedly announcing at one demonstration of the concept that it would 'do way altogether with existing methods of warfare.'[2] During the First World War, several countries pursued the development of what were then known as aerial torpedoes, early missiles based on radio-controlled biplanes. General Henry 'Hap' Arnold, an early proponent of missile and drone development during both world wars, said in 1945 that *'The next war may be fought by airplanes with no men in them at all.'*[3] Though many of these experiments proved largely unsuccessful, they served as the basis for additional development and testing during and after World War II and beyond.

The Cold War marked the first examples of large-scale operational use of reconnaissance drones, as well as the first tests of early UCAVs. In the 1960s, the US Navy developed a drone that could fit onboard destroyers and carry Mk 44 torpedoes or a nuclear depth charge for anti-submarine warfare missions.[4] The US Navy eventually produced 750 QH-50 Drone Anti-Submarine Helicopters (DASH), though the Navy was never able to track submarines well enough to engage them with the drone. In the early 1970s, the US Air Force studied the possibility of weaponizing the BQM-34 Firebee, a jet-powered drone that performed thousands of reconnaissance missions during the Vietnam War. Under a US Air Force contract awarded in 1971, Teledyne Ryan conducted several successful demonstrations of multi-mission drones, launching AGM-65 Maverick air-to-surface missiles and Mk 81 glide bombs at mock targets.[5] But the Air Force ultimately abandoned the project, postponing the first deployment of armed drones for another 30 years.

Around the turn of the century, the US Air Force, seeking to capitalize on deployments of unarmed RQ-1 Predator drones, launched a project to equip the Predator with air-to-surface Hellfire anti-tank missiles. Writing in May 2000, Air Force General John Jumper, then the commander of Air Combat Command, envisioned an expanded role for UCAVs in future conflict beyond reconnaissance and surveillance.[6] Adapting the missile to the Predator was no small task; the drone's wings had to be modified to withstand the weight of the missile rack and accommodate the force of the release, while the Predator's software needed to be amended to allow for strikes.[7] In January 2001, around seven months after launching the project, Air Force engineers conducted the first Predator missile test at China Lake test range, launching an inert Hellfire missile at a target tank. Some nine months later, in September 2001, the Air Force readied the armed Predator for its first operational mission over Afghanistan.

Over the next two decades, unmanned combat aerial vehicles would assume an increasingly prominent place in the conduct of war. The development and use of armed drones have proliferated widely beyond the United States, introducing new threats to militaries and innocent bystanders alike. The types of UCAVs and of drone munitions have also evolved since the Predator's early missile tests, becoming more diverse and multi-faceted than they were before.

This book documents the unmanned combat aerial vehicles that are in development or deployed, as well as the global operations in which they have been used. In doing so, it provides readers with a guidebook to navigating today's world of armed drones and drone munitions.

1   Everett, H. R., *Unmanned Systems of World Wars I and II* (The MIT Press, 2015), p247

2   Everett, H. R., *Unmanned Systems of World Wars I and II* (The MIT Press, 2015), p265

3   Kreps, S. and Zenko, M., 'The Next Drone Wars: Preparing for Proliferation', *Foreign Affairs* 93, No. 2 (2014) https://www.foreignaffairs.com/articles/2014-02-12/next-drone-wars

4   Armstrong, B., 'Unmanned Naval Warfare: Retrospect and Prospect', *Armed Forces Journal*, 20 December 2013, http://armedforcesjournal.com/unmanned-naval-warfare-retrospect-prospect/

5   Ehrhard, T. P., 'Air Force UAVs: The Secret History', Mitchell Institute, July 2010, https://apps.dtic.mil/sti/pdfs/ADA526045.pdf

6   Whittle, R., *Predator: The Secret Origins of the Drone Revolution* (New York, NY: Picador, 2015), p167

7   Whittle, R., *Predator: The Secret Origins of the Drone Revolution* (New York, NY: Picador, 2015), p175

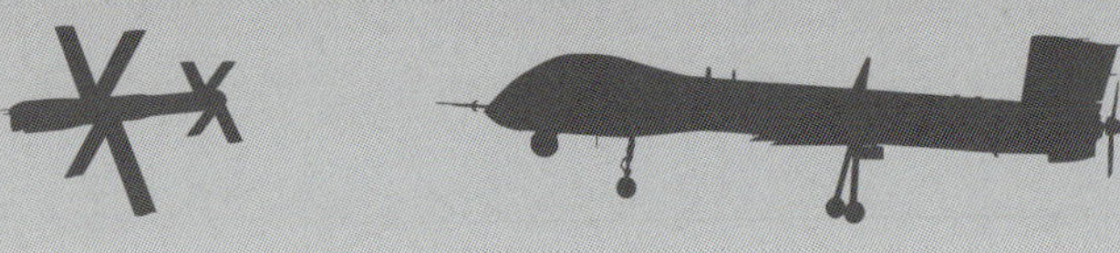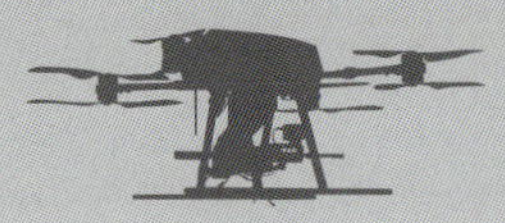

Acknowledgments

I would like to express my deep gratitude to my dear friend Arthur Holland Michel, whose thoughtfulness and integrity I will forever admire. My great thanks, too, to Thomas Keenan, who took a chance on the Center for the Study of the Drone and never faltered in his guidance and support.

I am supremely grateful to publisher Dr Heinz Berger and to the entire team at Harpia Publishing for their dedication, patience, and skill. I am also very grateful to fellow Harpia author Andreas Rupprecht, a peerless expert on Chinese military aircraft and a great help in this process. My thanks also goes to Joe Copalman, Georg Mader, Stijn Mitzer, Joost Oliemans, Dimitrije Ostojic, Igor Salinger, Angad Singh and Keyvan Tavakkoli for their photo contributions. Thank you all so very much.

Abbreviations

BLOS	beyond line of sight
Five Eyes	Australia, Canada, New Zealand, United Kingdom and USA
ft	Feet
GNA	Government of National Accord, in Libya
IDEX	International Defence Exhibition & Conference, Abu Dhabi, UAE
IRGC	Islamic Revolutionary Guard Corps
ISIS	Islamic State of Iraq and Syria
ISR	intelligence, surveillance and reconnaissance
kg	Kilogram
km/h	Kilometers per hour
kts	Knots
lb	Pounds
LMM	Lightweight Multirole Missile (Thales)
LNA	Libyan National Army
LOS	line of sight
mph	miles per hour
MTOW	maximum take-off weight
NATO	North Atlantic Treaty Organisation
PLA	People's Liberation Army, China
PLAN NA	People's Liberation Army Naval Aviation, China
RAAF	Royal Australian Air Force
RAF	Royal Air Force, United Kingdom
SDF	Sino Defence Forum
UAE	Unted Arab Emirates
UAV	unmanned aerial vehicle
UCAV	unmanned aerial combat vehicle
UN	United Nations
USAF	United States Air Force
USD	US Dollar
VTOL	vertical take-off and landing

Definitions

Different communities have their word – or acronym – of choice for describing uncrewed aircraft. In this book, the words 'drone'and 'UCAV' are used interchangeably, as are 'unmanned' and 'uninhabited' and 'uncrewed'. This book addresses two types of armed drones: UCAV and loitering munition. The latter is designed to be reusable, while the former is expendable.

This book does not include unarmed UAVs, of which there are many varieties, except when that system is known to have or is believed to be in the process of becoming capable of conducting strikes. This book also does not address historical armed platforms that are no longer in use, except when they appear in the descriptions of operations.

An unmanned combat aerial vehicle (UCAV) is a reusable aircraft designed to conduct kinetic operations. A UCAV is typically either a fixed-wing aircraft, as in one with wings that is designed for horizontal flight, or a rotary-wing aircraft, one that features one or more rotors and designed to take off and land vertically (VTOL). In theory, a UCAV could be a hybrid VTOL and fixed wing, though the only hybrids included in this book are loitering munitions. A UCAV is capable of being equipped with one or more munitions such as missiles and bombs. A fixed-wing UCAV typically carries munitions on pylons attached to the underside of each wing, while most rotary-wing UCAVs are equipped with munitions underneath the fuselage or on two outrigger pylons. Most fixed-wing UCAVs require a runway for take-off and landing, though some may be hand- or catapult-launched. Whether fixed- or rotary-wing, all UCAVs as of this writing require some level of human interaction to operate, usually in the form of a ground control station linked to the aircraft with radio or satellite communications.

A loitering munition is an expendable uncrewed aircraft that is designed to explode on impact with a target. For this reason, loitering munitions are similar to missiles, as they too carry explosive payloads. Unlike most missiles, however, loitering munitions are designed to circle over a target area, enabling the operator – or an onboard automated system – to identify opportunities for attack. Like UCAVs, loitering munitions come in a variety of types and sizes and are designed to be used by different communities. Some loitering munitions, like Israel's Harpy, are launched from truck-mounted launchers and intended to target adversary air defences and armoured vehicles. Others, like the US Switchblade, are launch from a small tube that can be carried in a soldier's backpack. Many loitering munitions bear a physical resemblance to unarmed fixed- and rotary-wing drones and, since any drone can be equipped with munitions, there are plenty of examples of cases in which drones have been turned into expendable munitions. The loitering munitions addressed in this book are purpose-built for military use.

Classifications

Each of the UCAVs in this book are assigned a classification ranging from I to III: Class I (light), Class II (medium) and Class III (heavy). These classifications are adapted from NATO Standardization Agreement 4670 – NATO's guidance for training drone operators – and are based on the technical characteristics of the aircraft such as maximum take-off weight (MTOW) and on the roles and missions of the aircraft.

Class I aircraft generally weigh less than 150kg (331lb), while Class II weigh between 150 and 600kg (331 and 1,323lb), and Class III over 600kg (1,323lb).

Table 1: Definition of Class I, II and III

Aircraft or loitering munition class		Typical characteristics
Class I	mini/micro/small	Max take-off weight less than 150kg (331lb) Hand- or rail-launched Operating altitude up to 1,524m (5,000ft)
Class II	tactical	Max take-off weight between 150 and 600kg (331 and 1,323lb) Rail- or runway-launched Operating altitude up to 5,486m (18,000ft)
Class III	strategic	Max take-off weight greater than 650 kg (1,433lb) Runway usually required for launch and recovery Operating altitude up to 19,812m (65,000ft)

The Songar, a Class I UCAV. (Asisguard)

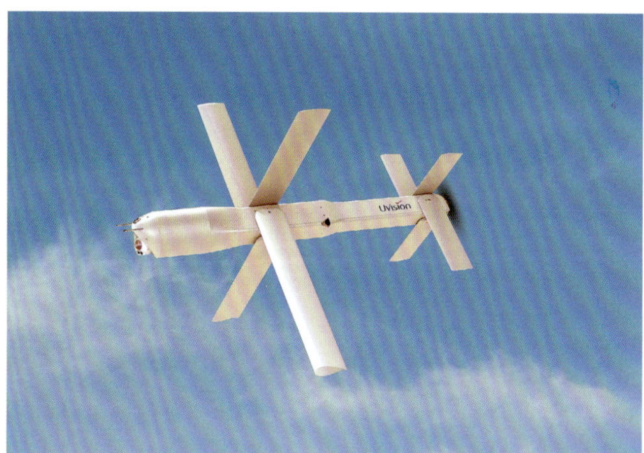

A Hero loitering munition. (UVision)

Some militaries have their own classification systems that may or may not correspond with NATO's guidance. The US, for example, separates drones into five groups according to weight, range and flight ceiling. Given the variety of UAVs, however, it is difficult for any classification system to capture the full range of capabilities.

The development of UCAVs, like many other aircraft, generally follow a predictable trend: the manufacturer presents a sub- or full-scale mock-up of the aircraft, the aircraft conducts its inaugural flight some months – or years – later, followed by weapons trials, and then acceptance tests by a particular user and eventually, deployment in exercises and operations. But, as with any emerging military technology, it can be difficult for an outside observer to know where exactly an aircraft is on that timeline based on publicly available information. And there are numerous examples of drones, not to mention aircraft in general, failing to progress to the next stage. For UCAVs, this challenge

An example of a Class II UCAV: the Iranian Mohajer-6 (Keyvan Tavakkoli)

The U.S. Air Force's MQ-9 is a typical Class III UCAV. (USAF/SSG Brian Ferguson)

is compounded by the fact that weaponization typically lags behind flight. Some aircraft, though they may have completed a weapons trial, may be in service only as an unarmed platform.

To parse the different stages of aircraft development and weaponization, each of the UCAVs and loitering munitions in this book are assigned a status rating ranging from A to C. An 'A' rating signifies that the aircraft is likely in operational service in some armed or unarmed capacity, a 'B' rating that the aircraft is undergoing testing, while a 'C' rating indicates that the aircraft probably remains conceptual and has yet to take to the air.

For UCAVs, a second rating is applied to the status of weaponization, with 'A' indicating that weapons are believed to have been integrated with the aircraft, 'B' that weapons trials have taken place, and 'C' that the weaponization of the aircraft remains conceptual and untested. For the purposes of this book, an A/A UCAV is one that is likely in service in some capacity as an armed platform. Of course, these subjective ratings may just as easily reflect the lack of available information as much as the presumed status of the aircraft. As such, they should be considered a shorthand for the purpose of the analysis in this book and not a statement on the actual condition of the aircraft.

Table 2: Definition of status A, B and C

Status	Aircraft or loitering munition	Munition
A	Aircraft or loitering munition appears to be in operational service either in an armed or unarmed capacity	Munition appears to have been integrated with a UCAV and used in active service
B	Aircraft or loitering munition appears to be undergoing flight trials but does not appear to be in operational use	Munition appears to have been test fired from a UCAV but does not appear to be in operational use with UCAVs
C	Aircraft or loitering munition remains conceptual and has yet to conduct tests based on available information	Munition has been exhibited – or is otherwise associated – with a UCAV but does not appear to have been tested

Trends in UCAVs

At least 96 unmanned combat aerial vehicles produced by 59 entities in 27 countries are currently under development or military service. Nearly three-quarters of the UCAVs in this book were unveiled since the start of 2015. Likewise, the number of countries developing UCAVs has doubled over the same period. Of the 96 UCAVs, one-quarter are believed to be in active military service as an armed platform; of the remainder, around half have conducted weapons trials and half remain untested as an armed platform, though some may be in service as an unarmed platform for reconnaissance and surveillance. At least 25 countries are believed to operate UCAVs, most of which were acquired from China, Turkey, or the United States.

Class I UCAVs comprise a small – but growing – subset of the uninhabited combat air vehicle market. This book features profiles on 18 Class I UCAVs produced by 13 entities in nine countries, all of which have been unveiled since 2015.

All but four Class I UCAVs in this book are rotary-wing systems and most have a flying time of less than one hour. Unlike loitering munitions, which are similar in size, Class I UCAVs are designed to be recoverable and equipped with weaponized payloads, often underneath the main fuselage or on outrigger pylons.

The Chinese-made Zhanfu H16-V12, for example, is said to be compatible with an assault rifle and grenade launchers. The Quadro-1400, a drone from Belarus, carries two RPG-26 rocket-propelled grenade launchers. While most weaponized payloads for Class I UCAVs consist of unguided rockets and bombs, a handful of manufacturers are working on integrating guided missiles. Spain's Aertec is developing the 5kg (11lb) A-Fox missile that it says will be compatible with its Tarsis-25 and Tasis-75 UCAVs. Two Class I UCAVs – China's Blowfish A2/3 and Turkey's Songar – are believed to be in military service, while the remainder are in some stage of development.

At least 20 Class II UCAVs produced by 17 entities in 11 countries are currently under development or in military service. Most Class II UCAVs are fixed-wing aircraft and based on existing short- and medium-range tactical ISR UAVs. For example, the Iranian Mohajer-6, South African Seeker 400 and US Nightwarden are armed variants of pre-existing uncrewed ISR aircraft, each of which can trace their origins back to the 1980s.

Table 3: Countries operating UCAVs

Country	Type	Origin	Introduction
Algeria	Caihong-3B, Caihong-4B, United-40	China, UAE	2018
Azerbaijan	Bayraktar-TB2	Turkey	2020
China	Wing Loong-1, Wing Loong-2	China	2011
Egypt	Wing Loong-1	China	2016
Ethiopia	Mohajer-6 (*reported*)	Iran	2021
France	MQ-9A Reaper	USA	2014
Indonesia	Caihong-4B	China	2020
Iran	Mohajer-6, Shahed-123, Shahed-129	Iran	2012
Iraq	Caihong-4B (inactive)	China	2015
Israel	Hermes-450 Heron	Israel	2000s
Kazakhstan	Wing Loong-1	China	2016
Libya (GNA)	Bayraktar-TB2	Turkey	2019
Myanmar	Caihong-3B	China	2014
Nigeria	Caihong-3B	China	2015
Pakistan	Caihong-3B, Caihong-4B, Wing Loong-2	China, China, China	2015
Qatar	Bayraktar-TB2	Turkey	2019
Russia	Forpost-R, Orion	Israel, Russia	2019
Saudi Arabia	Caihong-4B, Kayarel	China, Turkey	2014
Serbia	Caihong-92A	China	2020
South Africa	Seeker-400	South Africa	2018
Turkey	Akinci, Anka, Bayraktar-TB2	Turkey, Turkey, Turkey	2016
Turkmenistan	WJ-600A/D	China	2016
Ukraine	Bayraktar-TB2	Turkey	2019
United Arab Emirates	Seeker-400, Wing Loong-1, Wing Loong-2	South Africa, China	2016
United Kingdom	MQ-9A Reaper	USA	2007
United States	MQ-9A Reaper, MQ-1C Gray Eagle	USA	2001

A popular Chinese export: this CH-4B is operated by the Royal Saudi Air Force. (via Andreas Rupprecht)

Half of the Class II UCAVs in this book were unveiled prior to 2015. At least four Class II rotary-wing UCAVs are currently in service or under development. Class II UCAVs have on average a maximum take-off weight of approximately 400kg (882lb), payload capacity of 95kg (209lb) and flying time of 10 hours. Most Class II UCAVs are compatible with a variety of lightweight air-to-ground munitions. Some of these include China's TL-2 guided missile, France's Lightweight Multirole Missile (LMM), Iran's Ghaem-1 guided missile, Russia's KAB-20 guided bomb and South Africa's Denel P2 guided missile, all of which weigh less than 25kg (55lb). Nearly half of the Class II UCAVs in this book are believed to be military service as armed platforms, while the remainder are in some stage of development or in active service as unarmed platforms.

At least 58 Class III UCAVs produced by 35 entities are in some stage of development or military service. The number of countries producing or developing heavy-class UCAVs has grown from nine countries in 2015 to 22 countries today. Many of the heavy-class UCAVs in development, such as South Africa's MA380L and Ukraine's Sokil-300, are fixed-wing aircraft with a conventional design similar to that of the U.S.-made MQ-1 Predator, the archetypal armed drone. Five countries are developing heavy rotary-wing UCAVs, though none of these aircraft are in active service as armed platforms as of this writing. The UCAVs unveiled in the past six years are generally heavier and larger than those produced prior to 2015, resulting in a greater payload capacity and flight time. Manufacturers in China produce the greatest number and variety of UCAVs, with at least 19 heavy class UCAVs currently under development or in service. Seven countries – China, Iran, Israel, Russia, Turkey,

the United Arab Emirates, and the United States – produce Class III UCAVs that are currently believed to be in military service as armed platforms. That number is expected to grow in the coming years as countries like Indonesia, Pakistan, and Taiwan introduce domestically produced UCAVs. Still, the global market for heavy-class UCAV is expected to remain comprised of a relatively small number of entities – five countries have exported UCAVs to foreign military customers, and some only in limited quantities.

Table 4: Countries with UCAV acquisition programmes

Country	Type	Origin
Albania	Bayraktar-TB2 (*reported*)	Turkey
Australia	MQ-9B SkyGuardian	USA
Belgium	MQ-9B SkyGuardian	USA
Czech Republic	*Undetermined*	
India	MQ-9B SkyGuardian	USA
Iraq	Bayraktar-TB2	Turkey
Morocco	Bayraktar-TB2, MQ-9 Reaper	Turkey USA
Nigeria	Wing Loong-2	China
Poland	Bayraktar-TB2	Turkey
Taiwan	MQ-9B SkyGuardian	USA
Thailand	SkyScout X	Thailand
Tunisia	Anka-S	Turkey
Turkey	Akinci, Aksungur	Turkey
Turkmenistan	Bayraktar-TB2	Turkey
Ukraine	Bayraktar-TB2	Turkey
United Arab Emirates	MQ-9B SkyGuardian	USA
United Kingdom	MQ-9B SkyGuardian	USA

A variety of Chinese-made UCAV ordnance seen at Zhuhai Air Show 2018. (Harpia Publishing)

Trends in loitering munitions

At least 57 loitering munitions produced by 28 entities are in development or military service. More than three-quarters of the loitering munitions included in this book were unveiled in the six years since 2015. Likewise, the number of countries producing loitering munitions has grown from a half dozen to 15 during the same period. Of these 57 loitering munitions, 24 are believed to be in military service, while the remainder are in some stage of development. Thirteen loitering munitions are believed to have been exported to one or more foreign military customers. Around half of the loitering munitions in this book weigh between 2 and 20 kg (4.4 and 44lb) and half weigh between 20 and 150kg (44 and 331lb). Three loitering munitions weigh less than 2kg 4.4lb) and one has a maximum take-off weight of greater than 150kg (331lb). Several manufacturers offer multiple versions of the same loitering munition based on aircraft mass, payload capacity, and launch method. Poland's WB Group, for example, offers four versions of the Warmate loitering munition ranging from 4.5 to 30kg (9.9 to 66lb),

while Israel's UVision offers nine versions of the Hero loitering munition ranging from 2 to 125kg (4.4lb to 276lb). The loitering munitions included in this book have an average flight time of around one-and-a-half hours and range of 80 km (49.7 miles).

Unlike Class I UCAVs, which are similar in size and weight, loitering munitions are designed to be expendable and are typically equipped with an integrated warhead, which in some instances can be swapped in and out depending on the circumstance and intended target. Poland's MSP Inntech, for example, advertises five warheads compatible with the Warble loitering munition: high-explosive fragmentation, high-explosive anti-tank fragmentation, thermobaric, training, and inert. The 57 loitering munitions include 50 fixed-wing aircraft and six rotary-wing, as well as one hybrid vertical take-off and landing and fixed-wing drone. A variety of launch mechanisms exist to accommodate loitering munitions of varying size and weight. The smallest loitering munitions may be launched by hand or by a 40-milimeter grenade launcher. Other small fixed-wing loitering munitions are designed to be tube-launched, while

The standard UCAV weapon in the west: two AGM-114 Hellfire missiles, loaded on a MQ-9 Reaper from the 386th Air Expeditionary Wing at Ali Salem Air Base in Kuwait. (USAF/SSG Mozer Da Cunha)

larger loitering munitions require a pneumatic catapult or a vehicle-mounted canister launch system. Once largely a stand-alone system, loitering munitions are increasingly integrated into other crewed and uncrewed vehicles and aircraft. The U.S. Marine Corps, for example, is working on integrating loitering munitions into both uncrewed ground and surface vehicles.

Trends in munitions

Eighty munitions produced in 17 countries are associated with the UCAVs profiled in this book. These include air-to-ground missiles, bombs, glide bombs, and rockets, as well as loitering munitions, air-to-air missiles, anti-ship missiles, and grenades and mortars. Approximately half of the munitions catalogued in this book are believed to be in either active use with UCAVs or have been tested with a UCAV, while the remainder have been exhibited or are otherwise associated with UCAVs. The average weight of the muni-

tions that are in service or have been tested and for which specifications are available is approximately 67kg (147 lb).

Lightweight munitions are enabling a growing number of Class I and II uncrewed aircraft to be equipped with precision guided munitions.

Weapons like the Fury, a 6kg (13lb) guided bomb compatible with the Nightwarden, and China's TL-2, a 16kg (35lb) guided missile, are providing UAVs that once might have limited to ISR with a strike option. Thailand, for example, is working integrating the Thales LMM on the Sky Scout X. The Spanish Ministry of Defence is meanwhile funding a programme to develop lightweight rockets capable of being launched from a Class I UCAV like the Aertec Tarsis.[8]

The miniaturization of precision-guided munitions and availability of light- and medium-class UCAVs could offer smaller tactical units such as an army battalion or company an organic lethal UAV capability. Doing so would reduce the dependence of these units on long-range, strategic UCAVs and other aircraft that require substantial infrastructure and manpower to operate.

As UCAV manufacturers have tended to produce larger and heavier aircraft in recent years, the quantity and diversity of the munitions compatible with these aircraft has also increased. Whereas Turkey's Anka-B is compatible with a handful of lightweight anti-tank missiles, Turkish Aerospace Industries has promised that its successor, the Aksungur, will be compatible with 12 different munitions. One of the Aksungur's first weapons trials involved the KGK-82 guided bomb, which weighs around 15 times as much as the munitions compatible with the Anka-B. The munitions compatible with the latest UCAVs are also capable of flying further than previous versions. Turkey's Akinci, for example, is capable of being equipped with the MAM-T, a new long-range variant of the MAM-family that is four times heavier and has three times the range of the MAM-C and MAM-L munitions. The diversity of the munitions compatible with these aircraft is a sign of the evolving role of UCAVs on the battlefield and of the range of targets that UCAVs may encounter in the future. The U.S. Air Force has successfully conducted air-to-air tests involving the MQ-9 Reaper and AIM-9X Block 2 missile, part of a study of the ways in which Reapers may contribute to a conflict against a near-peer competitor.

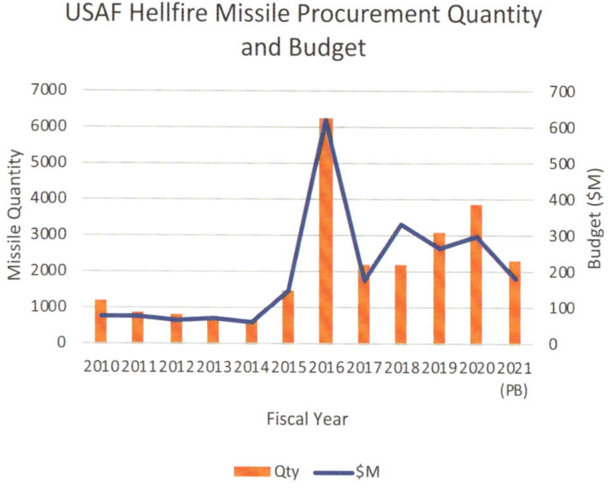

Trends in operations

At least 14 countries have deployed UCAVs in combat operations in 13 theatres in recent years. The U.S. became the first country to operate modern armed drones in 2001, followed by the Israel and UK in the mid-2000s. The number of countries with UCAV experience surged in 2015, when five countries – Iran, Iraq, Pakistan, Saudi Arabia and the UAE – are believed to have conducted their first combat operation. Nigeria and Turkey deployed UCAVs the following year. Azerbaijan, France and Russia in the last two years. It is possible that four other countries – Algeria, Egypt, Italy and Morocco – may also have deployed UCAVs, though the extent of these operations is not clear of this writing.

Table 5: Countries with UCAV operations experience (limited-substantial-extensive)

Country	Type	Year(s)
Azerbaijan	Limited	2020–intermittent
France	Substantial	2019–ongoing
Iran	Limited	2015–intermittent
Iraq	Limited	2015–2019
Israel	Extensive	2000s–ongoing
Libya (GNA)	Substantial	2019–ongoing
Nigeria	Limited	2015–intermittent
Pakistan	Limited	2015–2015
Russia	Limited	2019–2020
Saudi Arabia	Limited	2015–ongoing
Turkey	Extensive	2016–ongoing
UAE	Substantial	2015–ongoing
United Kingdom	Extensive	2007–ongoing
United States	Extensive	2001–ongoing

State-operated UCAVs are involved in combat operations in half of the 26 ongoing armed conflicts identified by the Council on Foreign Relations, a New York-based think tank, as of early 2021.[9]

UCAVs offer state actors multiple advantages in terms of cost, performance, and operational flexibility. As multirole aircraft, UCAVs are designed to be capable of conducting a range of missions on the battlefield from intelligence-gathering to strike. UCAVs can loiter over a particular area and conduct a post-strike battlefield damage assessment, shortening a targeting cycle that might otherwise involve multiple aircraft or ground forces. By virtue of being remotely piloted, UCAVs reduce the risk that crewmembers will be injured or killed, potentially allowing decisionmakers to take risks in deploying UCAVs that they might otherwise avoid. In terms of cost, UCAVs are generally cheaper to acquire and operate than crewed combat and ISR aircraft. The U.S. MQ-9 Reaper, for example, costs around USD 3,500 to operate per flying hour, while U.S. crewed combat aircraft typically exceed USD 20,000 per flying hour. Although figures vary, the procurement cost of a Bayraktar-TB2 and associated support equipment is estimated at between USD 4 and 5 million, considerably less than the USD 40 million crewed T129 attack helicopter Turkey lost in Syria in 2018.

These attributes have made UCAVs an attractive alternative to crewed combat aircraft in recent armed conflicts, particularly those involving asymmetric threats. Since the first armed drone mission in Afghanistan in 2001, UCAVs have been a feature of counterterrorism operations in Afghanistan, Iraq, Syria, and the Sahel. UCAVs have become a favourite tool of military interventions like that of the UAE in Libya, helping state actors avoid international scrutiny by providing a degree of plausible deniability. UCAVs have also featured prominently in domestic counterterrorism campaigns such as the ongoing conflict between Turkey and Kurdish groups in southeast Turkey. Even in inter-state conflicts such as those in Syria and Nagorno-Karabakh in 2020, in which UCAVs are at greatest risk of being the target of adversary defences, state actors have tended to rely on UCAVs while husbanding crewed combat aircraft for specialized tasks such as the suppression of adversary air.

UCAVs appear to have made a significant contribution to the total of number of strikes conducted by states in counterterrorism operations that involved a relatively small number of crewed and uncrewed combat aircraft. In France's Operation Barkhane in the Sahel, for example, in which the French Air Force has deployed seven crewed combat aircraft and three MQ-9 Reapers, the French UCAVs conducted around 45 per cent of all strikes in 2020, according to statements by French military officials. The French UCAVs also flew 40 per cent more hours than the crewed fighters. During the UK's UCAV deployments to Afghanistan between 2007 and 2014, the RAF's Reapers were responsible for 59 per cent of the flight hours and 51 per cent of the precision-guided munitions expended by fixed-wing combat aircraft, according to statistics released by the U.K. Ministry of Defence in 2015. During Operation Odyssey Lightning, a 2016 air campaign in Libya, three U.S. MQ-9 Reapers were responsible for approximately 60 per cent of airstrikes. By comparison, in larger operations, such as the US-led campaign against the self-styled Islamic State in Iraq and Syria, US MQ-1B Predators and MQ-9 Reapers had a smaller direct role in combat operations, averaging around 7 per cent of munitions expended between 2014 and 2017.[10]

Turkey's military campaigns in recent years have demonstrated the value of integrating UCAV operations with those of other air and ground assets. After losing more

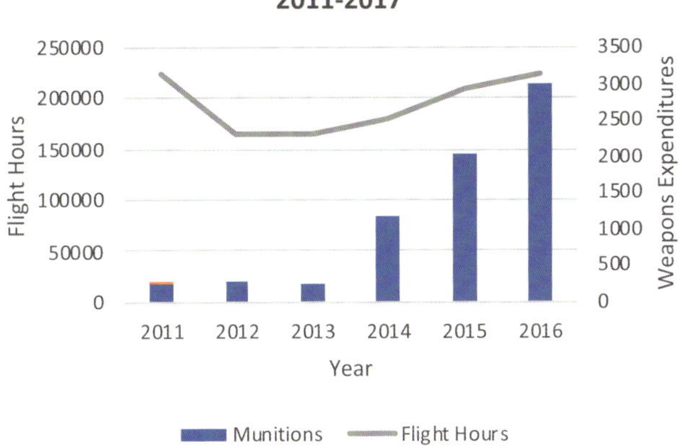

Source: '432d Remotely Piloted Aircraft Mission Brief'

than a dozen Bayraktar-TB2s to adversary aircraft and air defences in Libya in 2019, Turkey responded by deploying HAWK surface-to-air missiles and KORAL electronic warfare systems to Tripoli and Misrata. When Bayraktar-TB2s were downed in Syria's Idlib province in 2020, Turkish F-16 fighters launched strikes on an air base near Aleppo and downed two Syrian warplanes. And when Ankara provided Baku with Bayraktar-TB2s for operations in Nagorno-Karabakh, it also sent Turkish F-16s to Azerbaijan, which may have helped deter Armenian interceptors from encroaching on UCAV operations. The Turkish-Azerbaijani operations also involved drone decoys to distract and expose adversary air defences. In Turkish interventions in Syria and in Azerbaijani operations in Nagorno-Karabakh, UCAVs and loitering munitions were tasked in coordination with artillery, which acted as a force multiplier to target adversary supply lines. These examples of 'joint' operations illustrate how the integration of UCAVs and loitering munitions into a broader campaign plan can serve to alleviate some of the operational and technical limitations of armed drones when used in isolation.

It would be a mistake, however, to overestimate the effect that UCAVs have had on recent armed conflicts or to believe that UCAV experiences are universally shared. Some of these experiences have been limited and do not appear to have made a significant contribution to operational outcomes, nor to the direction of the conflict as a whole. This is certainly true of the short-lived deployments of Chinese-made drones by Iraq, Nigeria, and Pakistan, as well as Russia and Iran's limited operational experiences with UCAVs in Syria. Past operational experiences have highlighted the fact that a lack of trained personnel and replacement parts can negatively impact UCAV operations.

A deficit of trained personnel could limit the utility of UCAVs in future conflicts, even among those countries that appear to have relied heavily on UCAVs in recent deployments. In testimony before French legislators in 2020, military officials said that despite having enough Reaper aircraft, it has been difficult for France to train enough aircrews due to demands on existing operational crews.[11] The UK has also struggled with a shortage of Reaper aircrews, which observers have attributed to the stress experienced by many operators.[12] For the US Air Force, this has been a long-standing and persistent issue. As recently as 2020, over a decade after the MQ-9 Reaper's introduction, a US government report found that the Air Force continued to fall short of meeting staffing targets, even as the total number of USAF UCAV pilots has grown to exceed that of any other aircraft in the Air Force's inventory.[13] Even with trained personnel, inexperience could hinder future operations. It may prove difficult for Azerbaijan, for example, to replicate its success in Nagorno-Karabakh without the Turkish expertise they received last year. When acquiring UCAVs, it has not been uncommon for countries to underestimate the skill required to pilot the craft effectively and the manpower necessary to sustain operations.

The absence of a domestic industrial base and a reliance on foreign suppliers – nine of the 14 countries with some UCAV combat experience acquired aircraft from foreign manufacturers – is another constraint on UCAV operations. Iraq's fleet of 20 Caihong-4Bs have reportedly been inactive since 2019 due to of a lack of replacement parts from China, according to US investigators. And Nigeria has cited maintenance cost as a reason for why its Caihong-3Bs have languished. Nor are the countries that produce UCAVs domestically are immune to supply chain stressors. After Turkey's support for Azerbaijan in the 2020 Nagorno-Karabakh War, the Austrian and Canadian manufacturers of sensors and engines for the Bayraktar-TB2 announced they would suspend exports to Turkey, potentially endangering Baykar's ability to maintain its production lines. Turkey's efforts to produce domestic alternatives to these productions are ongoing.

Threats from adversary air defences , electronic warfare systems and aircraft continue to complicate UCAV operations. Of course, this is not a new phenomenon – there are numerous examples in the history of UAVs from Bosnia in 1995 to Georgia in 2005 and beyond of adversary air and anti-aircraft systems engaging UAVs. The proliferation of UCAVs has highlighted the vulnerability of these aircraft to defences. According to a database assembled by Drone

A MQ-9 Reaper ready to fly a training mission off Creech Air Force Base. These UCAVs are operated from all kind of locations worldwide. (USAF/A1C William Rio Rosado)

Wars UK, a London-based research organisation, more than four times as many military drones were downed in 2019 and 2020 by adversary actions as were downed in the previous 10 years combined.[14]

Poor operational planning, as well as the technical limitations of the systems themselves, can create inefficiencies in UCAV operations, eroding their effectiveness. The UAVs and UCAVs in the U.S.-led counter-ISIS campaign were overtasked, leading to a deficit in supply, according to researchers at the California-based RAND Corporation. Line-of-sight communications can limit the operational radius of UCAVs, as demonstrated by Turkey's experiences in Syria and Libya.

While emplacing ground relay stations to extend the range of the aircraft's communications, may alleviate this burden to some degree, doing so requires a friendly presence on the ground. Even for those countries with the technical and financial capacity to sustain the satellite bandwidth required for over the horizon command-and-control, aircraft basing is a key consideration given the slow speeds at which most UCAVs fly. At the outset of the counter-ISIS campaign, for example, UCAVs with the U.S.-led coalition expended one third of their maximum flight time transitioning from Kuwait to the area of operations in northern Iraq.

Conclusions and future trends

In the 20 years since the first UCAV strike in combat, armed drones have reshaped the modern battlespace. UCAVs and loitering munitions have grown in quantity and capability and demonstrated unique advantages and vulnerabilities in over a dozen operational theaters. Today, at least 40 countries are either developing or deploying UCAVs or loitering munitions, with others likely on the way.

What can we expect from armed drones over the next two decades? Recent trends suggest that the next generations of UCAVs will be more capable, loitering munitions more distributed and the roles and missions for armed drones more varied than those of the past two decades.

The development of so-called loyal wingman UCAVs could expand the types of combat missions for which drones are deployed. Unlike most traditional fixed-wing UCAVs, loyal wingman aircraft typically feature a low-observability airframe, jet propulsion and internal weapons bays. These characteristics are meant to enable UCAVs to match the survivability and speed of crewed combat aircraft, potentially enabling uncrewed aircraft to conduct more missions in contested airspaces than conventional UCAVs.

For example, the median maximum speed of the loyal wingman aircraft profiled in this book is more than 1.5 times that of all Class III UCAVs. In contrast to conventional UCAVs, many of the loyal wingman UCAVs under development today are designed to be low-cost and runway-independent, meaning that they launched in more austere environments than conventional UCAVs and lost without severe consequence.

At least seven countries – Australia, China, India, Russia, South Korea, Turkey and the U.S. – have unveiled around a dozen loyal wingman prototypes, all of which remain in some stage of development. These include Australia's Airpower Teaming System, India's Combat Air Teaming System Warrior and the U.S. XQ-58 Valkyrie, among others. Other countries such as Brazil, Finland, France, Germany, Italy, Japan, Spain and the U.K. are participating in or have launched their own efforts to develop loyal wingman UCAVs.

Loitering munitions are increasingly integrated into other crewed and uncrewed vehicles. Several manufacturers including AeroVironment, Area-I, IAI, STM and UVision are working on integrating loitering munitions into aircraft. The U.S. Area-I Altius-600, for example, can carry a 12kg (26lb) payload upwards of 400km (248 miles) and has been tested on board the U.S. Army's MQ-1C Gray Eagle and Air Force XQ-58 Valkyrie demonstrator. Israel Aerospace Industries, meanwhile, is collaborating with Korea Aerospace Industries on a project that may involve equipping crewed helicopters with loitering munitions. In the future, large UCAVs could act as carriers for swarms of air-launched effects that are deployed to target adversary air defences and other area-access area-denial platforms, a concept not dissimilar from that expressed by early drone visionaries.

On the ground, Germany's Rheinmetall and Estonia's Milrem have partnered with UAV manufacturers to integrate loitering munitions into crewed and uncrewed combat vehicles.

And at sea, the U.S. Navy and Marine Corps are working on equipping submarines and uncrewed surface vehicles with loitering munitions. Loitering munitions like the U.S. Raytheon Coyote are also the backbone of counter-drone systems designed to neutralize adversary drones. Future battlespaces could be saturated with loitering munitions of various kinds and capabilities.

A crewed fighter jet and its loyal wingmen – a familiar view in the future:? (Air Force Research Laboratory)

The ability of an uncrewed aircraft or loitering munition to operate independently of a human is an increasingly important element of UCAV research and development. The U.S. Skyborg research program, for example, seeks to develop a suite of hardware and software that will enable future loyal wingman aircraft to make decisions independent of the direct control of human teammates. Among loitering munitions, multiple manufacturers in Israel, Turkey and the UAE, among others, advertise systems that they say are equipped with autonomous capabilities or 'artificial intelligence', though the precise degree to which these systems are capable of independent action is not always clear. Emerging concepts of operations for uncrewed aircraft and changes in the operational environment are among the main drivers of the development of autonomy in uncrewed vehicles. Concepts such as human-machine teaming and collaborative swarms of UAVs and loitering munitions rely on aircraft capable of operating independently of human operators if they are to be deployed at scale. The growing risk to uncrewed aircraft – particularly those that rely on an assured communications link with an operator on the ground – from adversary air defences and electronic weapons is likewise encouraging manufacturers to pursue autonomous capabilities as a solution to operating in communications-denied environments. As countries like the U.S. and others transition from a focus on counterterrorism to preparing for a conflict with a peer competitor, autonomous capabilities are expected to become an increasingly important factor in UCAV and loitering munition design and use.

Although largely used to date as a ground-attack platform, UCAV and loitering munition operations could become a feature of the maritime domain. Unarmed uncrewed aircraft are already in service with more than a dozen navies around the world and performing a variety of missions at sea such as intelligence-gathering and target acquisition. Geopolitical concerns in the eastern Mediterranean, Black Sea, East and South China Seas, and elsewhere could encourage more militaries to seek a persistent, maritime-focused UCAV with offensive capabilities. The Ukrainian Navy became among the first to do so with its acquisition of strike-capable Turkish Bayraktar-TB2s in 2021. The U.S. military, meanwhile, has launched exercises to study the ways in which MQ-9 Reapers could be deployed to remote islands and engage in close air support strikes for Marines in an island attack scenario. Security threats in littoral zones from criminal and terrorist groups could likewise drive the acquisition of maritime-focused

UCAVs. The Houthi group in Yemen has demonstrated on multiple occasions how explosive-laden remote-controlled boats can threaten military and commercial vessels. UCAV manufacturers appear to be responding to the interest in uncrewed air-to-surface platforms. General Atomics Aeronautical Systems and Turkish Aerospace Industries have advertised heavy-class UCAVs equipped with anti-submarine payloads, while China has displayed UCAVs such as the Cloud Shadow alongside anti-ship missiles like the YJ-9E.

The marketplace for armed drones continues to grow at a rapid place, with more UCAVs and loitering munitions unveiled in 2021 than any other year to date. At the same time, the roles and missions for which armed drones may be used are also expanding as operational experiences, advanced munitions, emerging technologies, new concepts of operations, and geopolitical pressures push the boundaries of how militaries may employ these systems in future battlespaces. UCAVs and loitering munitions will test the vulnerability of other military assets such as tanks, as well as the preparedness of air defence systems and strategies. Far from making warfare unmanned, the proliferation of armed drones points to a future in which human soldiers and civilians are at greater risk, one for which every country ought to prepare.

8   CANCIO, F., 'El Ejercito quiere probar drones armados con cohetes guiados', *La Razon*, 18 November 2020, https://www.larazon.es/espana/20201118/kxqqge5lv5h7rizezbjdh-b4cpi.html

9   CFR STAFF, 'Global Conflict Tracker', Council on Foreign Relations, last updated 26 August 2021, https://www.cfr.org/global-conflict-tracker/?category=us

10   PAWLYK, O., 'F-15Es, A-10s Leading Air War Against ISIS', Military.com, 2 June 2017, https://www.military.com/daily-news/2017/06/02/f15es-a10s-leading-air-war-isis-stats-show.html/

11   ASSEMBLÉE NATIONALE, *Rapport D'information Déposé En Application De L'article 145 Du Règlement Par La Commission De La Défense Nationale Et Des Forces Armées En Conclusion Des Travaux D'une Mission D'information Sur L'opération Barkhane (No. 4089)*, Paris, 2021, accessed at: https://www.assemblee-nationale.fr/dyn/15/rapports/cion_def/l15b4089_rapport-information/

12   FISHER, L., 'Stress of killing from afar creates shortage of MoD drone operators', The Times, 13 January 2020, https://www.thetimes.co.uk/article/stress-of-killing-from-afar-creates-shortage-of-mod-drone-operators-0fnm36r6t

13   PAWLYK, O., 'The Air Force Risks a Drone Pilot Shortage, GAO Finds', Military.com, 26 June 2020, https://www.military.com/daily-news/2020/06/26/air-force-risks-drone-pilot-shortage-gao-finds.html

14   DroneWarsUK staff, 'Drone Crash Database', DroneWarsUK, accessed on 27 August 2021, https://dronewars.net/drone-crash-database/

ARMENIA

Pride Systems Copter and Drone

The Drone is a concept fixed-wing loitering munition developed by Pride Systems. The Drone features a variable-sweep wing design and can be tube-launched using a rocket booster. The Copter is a concept rotary-wing loitering munition developed by Pride Systems. It features a pentacopter design. Pride Systems unveiled a full-scale mock-up of the Drone and the Copter loitering munitions in February 2021 at the IDEX defence exhibition. On its website, Pride Systems describes both the Drone and the Copter as capable of conducting attacks autonomously based on a library of 'embedded video images of different targets in the guidance system'.

Pride Systems Copter. (Pride Systems)

Pride Systems Drone. (Pride Systems)

Specifications

	Copter
Class	Loitering munition
Status	C
Unveiled	2021
Operators	Armenia
Length	0.8m (2.6ft)
Width	0.6m (2.0ft)
Height	
Maximum take-off weight	5.0kg (11lb)
Speed (cruise)	150km/h (81kts)
(maximum)	200km/h (108kts)
Range	
Endurance	0.5 hours
Flight ceiling	
Payload	
Armament	

Specifications

	Drone
Class	Loitering munition
Status	C
Unveiled	2021
Operators	Armenia
Length	
Width	
Height	
Maximum take-off weight	5.0kg (11lb)
Speed (cruise)	150km/h (81kts)
(maximum)	200km/h (108kts)
Range	80km (50 miles)
Endurance	1.0 hours
Flight ceiling	
Payload	
Armament	

NERSISYAN, L., 'IDEX 2021 Armenian loitering munitions take a bow', *Shephard News*, 22 February 2021, https://www.shephardmedia.com/news/uv-online/idex-2021-armenian-loitering-munitions-take-bow/

PRIDE SYSTEMS, 'Products', accessed on 26 February 2021, https://pride.am/

AUSTRALIA

DefendTex Drone 40

The Drone40 is a rotary-wing loitering munition designed by DefendTex. The Drone40 has a cylindrical, canister fuselage and four collapsible rotors that extend when deployed. The Drone40 can be hand-launched or deployed using a grenade launcher. The Drone40 features a modular payload bay that can accommodate a 40mm grenade or payloads for intelligence-gathering, electronic warfare, smoke or flash grenades, or a laser designator. DefendTex unveiled the Drone40 at the Army Innovation Day 2016 defence exhibition. In January 2021, the British Army announced that it will equip the Royal Anglian regiment with the Drone40 during a deployment to Mali. In an interview with Australian Defence Magazine, a DefendTex spokesperson said that the company was working on 60mm, 81mm, and 155mm derivatives of the Drone40 for use by other Five Eyes military partners.

Specifications

Class	Loitering munition
Status	A
Unveiled	2016
Operators	United Kingdom
Length	120mm (0.4ft)
Width	120mm (0.4ft)
Height	
Maximum take-off weight	0.3kg (0.7lb)
Speed	
Range	20km (12 miles)
Endurance	0.2 hours
Flight ceiling	
Payload	0.11kg (0.2lb)
Armament	40mm grenade

DefendTex Drone40 (DefendTex)

'Drone40', Unmanned Aerial Vehicles, *DefendTex*, last updated 26 October 2020,
 https://www.defendtex.com/uav/

Levick, E., 'British Army takes Australian drone to Mali', *Australian Defence Magazine*, 28 January 2021,
 https://www.australiandefence.com.au/news/british-army-takes-australian-drone-to-mali

Boeing Airpower Teaming System

The Airpower Teaming System (ATS) is a demonstrator Class III fixed-wing UAV designed by Boeing in partnership with the Royal Australian Air Force (RAAF). The 'loyal wingman' system is designed to accompany manned fighters in combat operations, providing off-board sensing to the aircraft pilot. The ATS features a low-observable design with high-mounted, swept-back wings and V-tail vertical stabilizers. It was unveiled at the Australian International Airshow at Avalon in February 2019. Boeing began tests of subscale surrogate aircraft in November 2019. In December 2020, Boeing completed tests of five subscale aircraft flying autonomously as a team. The Airpower Teaming System conducted its maiden flight in February 2021. Australia has announced that it will acquire six of the aircraft for approximately USD 230 million. It is Australia's first domestically produced combat aircraft since World War II. It is likely that the ATS will be capable of carrying weapons, though the type and number remains unclear. Speaking in mid-2020, Air Marshal Hupfeld said that the RAAF was working on developing policies for deploying an armed loyal wingman aircraft.

Specifications

Class	III
Status	B/C
Unveiled	2019
Operators	
Length	11.7m (38ft)
Width (wingspan)	7.3m (24ft)
Height	
Maximum take-off weight	
Speed (max)	
Range	3,700km (2,300 miles)
Endurance	
Flight ceiling	
Payload	
Armament	

Boeing ATS. (Boeing)

'Boing completes teaming flights', *Australia Defence Magazine*, 3 December 2020,
    https://www.australiandefence.com.au/defence/air/boeing-completes-teaming-flights

Insinna, V., 'Australia makes another order for Boeing's Loyal Wingman drones after a successful first flight', *DefenseNews*, 3 March 2021,
    https://www.defensenews.com/air/2021/03/02/australia-makes-another-order-for-boeing-made-loyal-wingman-drones-after-a-successful-first-flight/

Pittaway, N., 'Boeing unveils 'loyal wingman' drone', *Defense News*, 27 February 2019,
    https://www.defensenews.com/digital-show-dailies/avalon/2019/02/27/boeing-unveils-loyal-wingman-drone/

Thorn, A., 'Chief of Air Force Talks Prepping for Loyal Wingman's first flight', *Australian Aviation*, 30 July 2020,
    https://australianaviation.com.au/2020/07/chief-of-air-force-talks-prepping-for-loyal-wingmans-first-flight/

AUSTRIA

Schiebel Camcopter S-100

The Camcopter S-100 is a Class II rotary-wing UAV developed by Schiebel. The S-100 is an improved version of the Camcopter 5.1. Schiebel unveiled the Camcopter S-100 at the 2005 IDEX defence exhibition. At the 2008 Farnborough Airshow, Schiebel displayed a Camcopter S-100 fitted with a 13kg (29lb) Thales Lightweight Multirole Missile (LMM) and announced that the S-100 had already conducted weapons trials. Today, however, the S-100 is widely used in an unarmed capacity for reconnaissance and surveillance. The S-100 is in active military service with at least 10 countries, though it does appear as of this writing that these countries operate the armed variant of the S-100.

Specifications

Class	II
Status	A/B
Unveiled	2005
Operators	None as UCAV
Length (main rotor diameter)	3.1m (10.2ft)
Width	3.4m (11.2ft)
Height	
Maximum take-off weight	200kg (440lb)
Speed	
Range	200km (124 miles)
Endurance	6 hours
Flight ceiling	
Payload	50kg (110lb)
Armament	LMM

Schiebel S-100 Camcopter (Milborne Systems)

Hambling, D., 'New Killer Drones Invade Airshow', *Wired*, 23 July 2008,
https://www.wired.com/2008/07/new-killer-dr-1/

Schiebel, 'Schiebel Introduces Next Generation UAV at IDEX 2005', news release, February 2005,
https://schiebel.net/wp-content/uploads/2015/07/2005-02_SCHIEBEL_INTRODUCES_NEXT-GENERATION_UAV_AT_IDEX_2005.pdf

BELARUS

Display Design Bureau Quadro-1400 and Loitering Tube

The Quadro-1400 (Квадро-1400) is a Class I rotary-wing UCAV developed by the Display Design Bureau. The Loitering Tube (Барражирующая труба) is a smaller variant of the Quadro-1400 that is also produced by Display Design Bureau. The Quadro-1400 is capable of equipping with two RPG-26 rocket-propelled grenade launchers, while the Loitering Tube can carry a single launcher fitted to the underside of the aircraft. Both drones were unveiled in early 2020, when the Display Design Bureau released video footage to Belarusian media outlets showing the Quadro-1400 and Loitering Tube conducting firing tests. The Belarusian military displayed both aircraft at a military parade in May 2020 commemorating the 75th anniversary of the end of World War II. The Display Design Bureau also exhibited the Quadro-1400 at Russia's Army 2020 defence exhibition in August. Also in August, Goskomvoenprom, the Belarusian state arms directorate, released video footage of the Loitering Tube conducting weapons trials with the PTAB-2.5 and PTAB-10.5 miniature unguided bombs. Both aircraft are intended to be used against tanks and other armoured vehicles.

Quadro-1400. (National Academy of Sciences of Belarus)

Loitering Tube. (National Academy of Sciences of Belarus)

Specifications

	Quadro-1400
Class	I
Status	B/B
Unveiled	2020
Operators	
Length	1.45m (4.8ft)
Width	1.45m (4.8ft)
Height	0.65m (2.0ft)
Maximum take-off weight	35kg (77lb)
Speed (maximum)	54km/h (29.2kts)
Range (operational)	5km (3.1 miles)
Endurance	0.5 hours
Flight ceiling	
Payload	
Armament	RPG-26

Specifications

	Loitering Tube
Class	I
Status	B/B
Unveiled	2020
Operators	
Length	0.7m (2.3ft)
Width	0.7m (2.3ft)
Height	0.34m (1.0ft)
Maximum take-off weight	11kg (34lb)
Speed (maximum)	65km/h (35.1kts)
Range (operational)	2km (1.24 miles)
Endurance	0.25 hours
Flight ceiling	
Payload	
Armament	Miniature bomb

KB Unmanned Helicopters Hunter

The Hunter is a Class III rotary-wing UCAV developed by Unmanned Helicopters Design Bureau, a Minsk-based company. According to the manufacturer, the Hunter can be equipped with a 7.62mm Kalashnikov tank machine gun and up to 500 rounds of ammunition; two rocket pods with a capacity of four unguided rockets apiece; or up 16 2.5kg (5.5lb) anti-tank miniature bombs. It was unveiled at the 2021 MILEX defence exhibition. Video released by the manufacturer appears to show the Hunter conducting flight tests.

Specifications

Class	III
Status	B/C
Unveiled	2021
Operators	
Length	8.3m (27ft)
Width (wingspan)	7.0m (23ft)
Height	
Maximum take-off weight	750kg (1,650lb)
Speed (max)	180km/h (97.2kts)
Range	
Endurance	9 hours
Flight ceiling	3,500m (11,500ft)
Payload	200kg (440lb)
Armament	Light machine gun, unguided rockets, miniature bombs

KB Unmanned Helicopters Hunter.
(State Military-Industrial Committee of the Republic of Belarus)

'Разведывательно-ударный беспилотный вертолет Hunter представлен на выставке MILEX', *Belta*, 23 June 2021, https://www.belta.by/photonews/view/razvedyvatelno-udarnyj-bespilotnyj-vertolet-hunter-predstavlen-na-vystavke-milex-25368/

INDELAUAV (@indelauav), 'uav helicopter Hunter 2021', YouTube, 23 June 2021, https://www.youtube.com/watch?v=DDQWhtUib-8

ATH: новости Беларуси и мира, 'Новые разработки белорусской военной техники. Панорама', YouTube, 31 January 2020, https://www.youtube.com/watch?v=CE49dPt9ap8

ATHERTON, K. D., 'For Belarus, two rockets on a drone are better than one', *C4ISRNET*, 19 February 2020, https://www.c4isrnet.com/unmanned/2020/02/20/for-belarus-two-rockets-on-a-drone-are-better-than-one/

BAZZOLO, G. A., 'Armed drones in service with Belarus army', *Defence Blog*, 13 July 2018, https://defence-blog.com/news/armed-drones-service-belarus-army.html

BELTECHEXPORT, 'BUREVESTNIK MB' LONG-RANGE UNMANNED AERIAL SPECIAL-PURPOSE SYSTEM', accessed on 14 March 2021, https://bte.by/en/katalog/aviatsionnaya-tekhnika/vertolety/bespilotnyy-aviatsionnyy-kompleks-dalnego-deystviya-spetsialnogo-naznacheniya-burevestnik-mb.html

BOSERMAN, M., 'Гранатометом с неба. Белорусская «шайтан-труба» Квадро-1400 на Армии-2020', *Nauka Technika*, 29 August 2020, https://naukatehnika.com/granatometom-s-neba-belorusskaya-kvadro-1400.html

PTICHKIN, S., 'Летать так летать: Белорусские беспилотники произвели фурор на форуме 'Армия-2016', *RG.RU*, 14 September 2016, https://rg.ru/2016/09/14/belorusskie-bespilotniki-proizveli-furor-na-forume-armiia-2016.html

'Беларусь покажет в Абу-Даби беспилотник с дронами-камикадзе', *RG.RU*, 14 February 2018, https://rg.ru/2018/02/14/belarus-pokazhet-v-abu-dabi-bespilotnik-s-dronami-kamikadze.html

Госкомвоенпром РБ, 'сбросы ПТАБ с беспилотника', YouTube, 13 August 2020, https://www.youtube.com/watch?v=OBZwlJIeD7w

SPC Burevestnik-MB

The Burevestnik-MB (БУРЕВЕСТНИК-МБ) is a Class II fixed-wing UCAV developed by the Research and Production Centre of Multifunctional Unmanned Systems, a division the National Academy of Sciences of Belarus. The Burevestnik-MB features mid-mounted wings, a rounded fuselage, and twin booms with a low-mounted horizontal stabilizer. It has a pusher propeller configuration and tricycle landing gear. Work on the Burevestnik-MB reportedly began in 2011. It was unveiled at the Army-2016 defence exhibition. At the Army-2016 and MILEX-2017 defence exhibitions, the Burevestnik-MB was displayed alongside an assortment of munitions, including an unspecified bomb, an unnamed loitering munition, and S-5 unguided rockets. In a July 2017 parade celebrating Belarus Independence Day, the Belarusian Armed Forces displayed two truck-mounted Burevestnik-MBs fitted with these unnamed loitering munitions, as well as bombs. A similar display was repeated at other events, such as the UMEX 2018 defence exhibition. According to BelTechExport, the Belarusian trade agency, the Burevestnik-MB's compatible armament includes 'high-precise gliding UAVs' – the unnamed loitering munitions – and S-5 and S-8 rockets. These unnamed gliding loitering munitions reportedly weight around 26kg (57lb) and have a payload capacity of 10kg (22lb). However, the extent to which the Burevestnik-MB has conducted weapons trials or operational deployments is unclear. In April 2020, Vladimir Gusakov, the chairman of the National Academy of Sciences, announced that Belarus would establish a UAV manufacturing centre in Egypt to produce drones, which could include the Burevestnik-MB.

Both, the Burevestnik-MB and the below mentioned loitering munition have been seen at the KADEX 2018 defence expo.
(Belarms Catalouge 2018)

Specifications

	Burevestnik-MB
Class	II
Status	B/C
Unveiled	2016
Operators	
Length	
Width	
Height	
Maximum take-off weight	400kg (880lb)
Speed (cruise)	150km/h (81kts)
(maximum)	220km/h (118.8kts)
Range (operational)	300km (186 miles)
Endurance	10 hours
Flight ceiling	5,000m (16,400ft)
Payload	60kg (132lb)
Armament	S-5, S-8, unnamed loitering munition

Specifications

	Loitering munition
Class	Loitering munition
Status	C
Unveiled	2017
Operators	
Length	
Width	
Height	
Maximum take-off weight	25kg (55lb)
Speed	200km/h (108kts)
Range (operational)	36km (22 miles)
Endurance	
Flight ceiling	
Payload	10kg (22lb)
Armament	

CHINA

AVIC AV500B

The AV500B is a Class II rotary-wing UCAV developed by AVIC. It was unveiled at the 2016 Zhuhai Airshow. The AV500B is the armed reconnaissance variant of the AV500 family of the rotary-wing UAVs. At the time of unveiling, it was China's first strike-capable unmanned rotary-wing aircraft. The AV500B has two outrigger pylons. In November 2018, Chinese media reported that the AV500B had carried out a successful weapons trial, destroying a target with an FT-8D air-to-ground guided missile at a range of 4.5km (2.8 miles). The U8EW is an exportable, armed variant of the AV500B. It was displayed at the 2018 Singapore Air Show and at the 2019 Dubai Air Show alongside the TL-8, a 16kg (35lb) guided missile.

Specifications

Class	II
Status	A/A
Unveiled	2016
Operators	China
Length	7.2m (23.6ft)
Width (wingspan)	
Height	
Maximum take-off weight	470kg (1,035lb)
Speed (maximum)	170km/h (105kts)
Range	200km (125 miles)
Endurance	8 hours
Flight ceiling	
Payload	160kg (350lb)
Armament	FT–8D, TL–2

An AV500W shown as a mock uo at the Zhuhai 2016 Airshow. Two years later in 2018 the same UAV reappeared in slightly revised form. (PLA)

Air Recognition, 'AVIC AV500W UAV completes FT-8D missile live-fire test', *Air Recognition*, 5 November 2018, http://www.airrecognition.com/index.php/archive-world-worldwide-news-air-force-aviation-aerospace-air-military-defence-industry/global-defence-security-news/global-news-2018/november/4602-avic-av500w-uav-completes-ft-8d-missile-live-fire-test.html

Donald, D., 'China Shows Armed Rotary-Wing UAS', *Aviation International Online*, 18 November 2019, https://www.ainonline.com/aviation-news/defence/2019-11-18/china-shows-armed-rotary-wing-uas

Kenhmann, H., 'AV500W: l'Arabie Saoudite s'intéresse au drone VTOL armé chinois', East Pendulum (blog), 9 December 2016, http://www.eastpendulum.com/av500w-helidrone-arme-interesse-arabie-saoudite

AVIC Cloud Shadow

The Cloud Shadow (云影), also known as the Wing Loong-10 or Wind Shadow, is a demonstrator Class III fixed-wing UCAV developed by AVIC. A full-scale mock-up of the Cloud Shadow was unveiled at China's Zhuhai Airshow 2016. The Cloud Shadow features a sleek fuselage design, bulbous nose, and V-tail. It is designed to carry out high-altitude surveillance and strike missions. AVIC displayed another mock-up of the Cloud Shadow at the Nanchang Aviation Conference in October 2020 that featured a ZFTX Aeroengine ZF850 turbojet engine. It was displayed alongside a variety of munitions, including the BA-21, GB-4, LS-6100, BBM2, and BBM3 guided bombs, FT-7 glide bomb, YJ-9E anti-ship missile, and

BA-7 and AG-300/M guided missiles. However, pending confirmation of weapons trials, the final loadout may or may not include these munitions. The Cloud Shadow has three hardpoints under each wing. At the time of unveiling in 2016, AVIC announced that it would also be available for export to international customers. In August 2020, AVIC announced that the Cloud Shadow had conducted a high-altitude flight in the southern Hainan province to monitor meteorological conditions. During the flight, AVIC reported that the Cloud Shadow cruised at an altitude of 10,000m (32,808ft) and deployed aerosondes to monitor a tropical storm.

Specifications

Class	III
Status	B/C
Unveiled	2016
Operators	
Length	9.0m (29.5ft)
Width (wingspan)	17.0m (56ft)
Height	
Maximum take-off weight	3,000kg (6,600lb)
Speed (maximum)	
Range (LOS)	290km (180 miles)
Endurance	
Flight ceiling	
Payload	400kg (880lb)
Armament	

East Pendulum (@HenriKenhmann), 'La version d'attaque au sol du #drone Cloud Shadow est présentée actuellement au Meeting aérien 2020 de Nanchang', Twitter, 1 November 2020, https://twitter.com/HenriKenhmann/status/1322923881955430400

Minnick, W., 'Zhuhai 2016: China reveals Cloud Shadow UAV', *Shephard Media*, 31 October 2016, https://www.shephardmedia.com/news/uv-online/zhuhai-2016-china-reveals-cloud-shadow-uav-export/

Wong, K., 'China reveals rare glimpse of Wind Shadow UAV', *Janes*, 4 August 2020, https://www.janes.com/defence-news/news-detail/china-reveals-rare-glimpse-of-wind-shadow-uav

The Cloud Shadow, seen at Zhuhai in 2018, has been renamed Wing Loong 10 for the export market since mid–2020. Although it is in service for meteorological survey, the armed variant apparently still lacks a customer. (Harpia Publishing)

AVIC GJ-11 Lijian

The GJ-11 Lijian, or Sharp Sword, is a concept Class III fixed-wing UCAV developed by the Aviation Industry Corporation of China (AVIC). It features a tailless flying-wing design with a rear-mounted WS-13 turbofan engine. Photographs of the GJ-11 at the 2021 Zhuhai Airshow show that the aircraft has two internal weapons bays, each with a capacity of up to four munitions. Official specifications for the GJ-11 are not available as of this writing, though it is estimated to have a wingspan of around 14m (46ft). The GJ-11 is one of several designs for stealthy combat UAVs that AVIC developed under the 601-S programme, which has been underway since at least 2013. The People's Liberation Army displayed the GJ-11 in a 2019 military parade celebrating the 70th anniversary of the People's Republic of China, suggesting that it may already be in military service. However, it is not clear as of this writing that the GJ-11 has conducted either flight tests or weapons trials.

Specifications

Class	III
Status	C/C
Unveiled	2019
Operators	
Length	
Width (wingspan)	
Height	
Maximum take-off weight	
Speed (maximum)	
Range	
Endurance	
Flight ceiling	
Payload	
Armament	

A GJ-11 Sharp Sword with its bomb bays open at the Zhuhai airshow in 2021. (Goneless)

D'Urso, S., 'China Exhibits New Sharp Sword UCAV During Military Parade for PRC's 70th Anniversary', *The Aviationist*, 1 October 2019, https://theaviationist.com/2019/10/01/china-exhibits-new-sharp-sword-ucav-during-military-parade-for-prcs-70th-anniversary/

AVIC Wing Loong-1

The Wing Loong-1 a Class III fixed-wing UCAV developed by the Chengdu Aircraft Industry Group, a division of AVIC. The Wing Loong-1, which is also known as the Pterodactyl-1 or GJ-1, features a V-tail and mid-mounted wings. AVIC displayed a subscale model of the Wing Loong-1 at the 2010 Zhuhai Airshow, though it may already have conducted its maiden flight by then. AVIC unveiled a full-scale mock-up at the 2012 Zhuhai Airshow. At the 2014 Zhuhai Air Show, AVIC displayed a Wing Loong-1 in the livery of the People's Liberation Army Air Force, indicating that it was in service with the Chinese military. In September 2015, the People's Liberation Army displayed two Wing Loong-1s at the Victory Day parade in Beijing.

At the 2012 Zhuhai Air Show, AVIC displayed the Wing Loong-1 with the Blue Arrow-7 air-to-ground guided missile, an export-oriented variant of the AKD-10 missile, as well as the LS-6(50), YZ-102, and YZ-200 guided bombs. At the 2016 Zhuhai Airshow, AVIC displayed the Wing Loong-1 with a more extensive array of munitions than it had in 2012. These included the 11kg (24lb) CM-502KG, 16kg (35lb) TL-2, YJ-9E, 13kg (29lb) GAM-101A/B, and AG-300L/M air-to-ground guided missiles; the 25kg (55lb) FT-10 guided bomb and 50kg (110lb) FT-9 guided bombs; as well as the LS-6, YZ-102, YZ-200, and

Blue Arrow-7 that were displayed with the Wing Loong-1 previously. Of these, AVIC has published video footage of Wing Loong-1 weapons trials with the Blue Arrow-7 and the LS-6.

The Wing Loong-1 is in military service with China, Egypt, Kazakhstan, and the United Arab Emirates. In 2016, Pakistan acquired two Wing Loong-1s for test and evaluation, though that programme appears to have ended. The UAE has deployed the Wing Loong-1 to Libya and Yemen, while Egypt has used the Wing Loong-1 to support counterterrorism operations on the Sinai Peninsula and to monitor its western border with Libya. The Wing Loong-1 is in service with the People's Liberation Army Air Force's 178th UAV Brigade based at Malan Air Base. In March 2021, Kazakhstan's Ministry of Defence released a promotional video that appears to show its Wing Loong-1 conducting weapons trials with Blue Arrow-7 missiles, the first indication that Kazakhstan's Wing Loong-1s are armed.

Specifications

Class	III
Status	A/A
Unveiled	2010
Operators	China, Egypt, Kazakhstan, UAE
Length	9.0m (29.5ft)
Width (wingspan)	14.0m (46ft)
Height	
Maximum take-off weight	1,150kg (2,530lb)
Speed (maximum)	280km/h (151.2 kts)
Range	200km (124 miles)
Endurance	20 hours
Flight ceiling	
Payload	200kg (440lb)
Armament	Blue Arrow-7, LS-6 (50)

GETTINGER, D., 'The Drone Databook', *The Centre for the Study of the Drone at Bard College*, September 2019, dronecenter.bard.edu/databook

PECK, M., 'Watch China's Wing Loong UAV test weapons', *C4ISRNET*, 24 August 2015, https://www.c4isrnet.com/unmanned/2015/08/24/watch-china-s-wing-loong-uav-test-weapons/

SHEPHARD NEWS TEAM, 'Formation of Pterosaur UAVs soars in China', 16 January 2015, https://www.shephardmedia.com/news/uv-online/formation-pterosaur-uavs-soars-china/

WALDRON, G., 'Zhuhai 10: PICTURES: China reveals armed UAV designs', *FlightGlobal*, 25 November 2010, https://www.flightglobal.com/zhuhai10-pictures-china-reveals-armed-uav-designs/97102.article

XINHUA NEWS AGENCY, 'CHINA-ZHUHAI-AVIATION-EXHIBITION-DRONE (CN)', *gettyimages*, 3 November 2016, https://www.gettyimages.com/detail/news-photo/nov-3-2016-a-wing-loong-i-unmanned-aerial-vehicle-is-news-photo/620895292

A Wing Loong I, known in PLAAF service as the GJ-1, assigned to the Western Theater Command's UAV brigade. (Air Victoria)

AVIC Wing Loong-2

The Wing Loong-2 is a Class III fixed-wing UCAV developed by the Chengdu Aircraft Industry Group, a division of AVIC. Also known as the Pterodactyl-2 or GJ-2, the Wing Loong-2 is larger and more capable than its predecessor. Like the Wing Loong-1, the Wing Loong-2 has a bulbous nose, mid-mounted wings, and a V-tail. The Wing Loong-2 has six hardpoints and has more than twice the external payload capacity of the Wing Loong-1. AVIC unveiled a full-scale mock-up of the Wing Loong-2 at the 2015 Beijing Air Show. The Wing Loong-2 conducted its maiden flight in 2017. The People's Liberation Army displayed a Wing Loong-2 equipped with various unspecified munitions at the 70th founding anniversary parade in October 2019.

Like the Wing Loong-1, AVIC has also publicly displayed the Wing Loong-2 with a variety of air-to-surface missiles and bombs. At the 2017 Paris Air Show, the Wing Loong-2 was accompanied by the Blue Arrow-7, AGM-300M, and CM-502 air-to-surface missiles, the YJ-9E anti-ship missile, and the FT-9, FT-10, and LS-6 guided bombs. AVIC presented a similar array at the 2018 Zhuhai Air Show, adding the TY-90 air-to-air missile. A 2018 analysis by *Janes* found that other compatible weapons may include the Blue Arrow-9 and Blue Arrow-21 missiles and the GB-3 and GB-7 laser-guided bombs. In early 2018, AVIC published video footage of a Wing Loong-2 weapons trial with the Blue Arrow-7 and TL-2 air-to-ground missiles, the 100kg (220lb) LS-6(100) guided bomb, and what appeared to be the TY-90 air-to-air missile. Publicly available photos of the Wing Loong-2 show that it with a total of either four or six external hardpoints. In a January 2021 photo of a ceremony marking the handover of the 50th Wing Loong-2, the aircraft is pictured with three hardpoints under each wing, each of which carried multiple missiles.

The Wing Loong-2 is in military service with China, the United Arab Emirates, and the Libyan National Army, a UAE proxy force under the command of Khalifa Haftar. In late 2020, Nigeria announced that it had ordered several Wing Loong-2, becoming the third country to operate the aircraft. The UAE has deployed Wing Loong-2 equipped with Blue Arrow-7 missiles in support of the Libyan National Army. In a December 2019 report, the United Nations Security Council has documented the use of Wing Loong-2 and Blue Arrow-7 missiles in Libya.

Specifications

Class	III
Status	A/A
Unveiled	2015
Operators	China, Nigeria, UAE
Length	11.0m (36ft)
Width (wingspan)	20.5m (67ft)
Height	4.1m (13ft)
Maximum take-off weight	4,200kg (9,240lb)
Speed (maximum)	370km/h (199.8kts)
Range (operational)	200km (124 miles)
Endurance	20 hours
Flight ceiling	9,000m (39,528ft)
Payload	480kg (1,056lb)
Armament	Blue Arrow-7, LS-6, TL-2, TY-90

Arthur, G., 'China takes wraps off Wing Loong 2', *Shephard News*, 21 September 2015, https://www.shephardmedia.com/news/uv-online/china-takes-wraps-wing-loong-2/

Carey, B., 'Under Trump administration, US reviews drone export policy', *Aviation International Online*, 31 August 2017, https://www.ainonline.com/aviation-news/defence/2017-08-31/under-trump-administration-us-reviews-drone-export-policy

Imaginechina, 'A Wing Loong II unmanned aerial vehicle (UAV), or drone, of Aviation Industry Corporation of China (AVIC)', *Alamy*, https://www.alamy.com/a-wing-loong-ii-unmanned-aerial-vehicle-uav-or-drone-of-aviation-industry-corporation-of-china-avic-is-on-display-during-the-12th-china-internat-image261826784.html?pv=1&stamp=2&imageid=34C420DA-6297-4AC9-98AD

Kenhmann, H., 'Drone : Chengdu livre les premiers lots de Wing Loong II aux clients', East Pendulum (blog), 2 January 2018, http://www.eastpendulum.com/drone-chengdu-livre-premiers-lots-de-wing-loong-ii-aux-clients

Nurkin, T., 'China's Advanced Weapons Systems', *Jane's*, 12 May 2018, https://www.uscc.gov/sites/default/files/Research/Jane's%20by%20IHS%20Markit_China's%20Advanced%20Weapons%20Systems.pdf

Tomkins, R., 'Chinese drone Wing Loong II conducts maiden flight', *UPI*, 28 February 2017, https://www.upi.com/Defence-News/2017/02/28/Chinese-drone-Wing-Loong-II-conducts-maiden-flight/2881488295445/

The Wing Loong II — known as the GJ-2 in PLA service — during a demonstration flight at the 2021 Zhuhai Airshow. (SDF)

AVIC Wing Loong ID

The Wing Loong ID (翼龙ID) is a demonstrator Class III fixed-wing system developed by Aviation Industry Corporation of China (AVIC). A scale model of the Wing Loong ID's design, which is based on the Wing Loong 1, was unveiled at the 2016 Zhuhai Airshow. A prototype aircraft made its maiden flight at Anshun Huangguoshu Airport on 23 December 2018. The ID is intended for the export market and is expected to make its first deliveries in the coming years.

Specifications

Class	III
Status	B/B
Unveiled	2016
Operators	
Length	8.7m (28.5ft)
Width (wingspan)	17.6m (58ft)
Height	3.2m (10ft)
Maximum take-off weight	1,500kg (3,300lb)
Speed (maximum)	280km/h (151.2kts)
Range (LOS)	200km (124 miles)
Endurance	35 hours
Flight ceiling	7,500m (24,600ft)
Payload (external)	400kg (880lb)
Armament	

Global Security,
https://www.globalsecurity.org/military/world/china/images/wing-loong-image12.jpg

Waldron, G., 'AVIC's Wing Loong ID UAV takes flight', *FlightGlobal*, 26 December 2018,
https://www.flightglobal.com/news/articles/avics-wing-loong-id-uav-takes-flight-454655/

Wong, K., 'China's AVIC aims to roll out Wing Loong I-D in 2018', *Jane's International Defence Review*, 29 January 2018,
https://www.janes.com/article/77414/china-s-avic-aims-to-roll-out-wing-loong-i-d-in-2018

The so far sole Wing Loong ID demonstrator was on display at Zhuhai in 2018. (Harpia Publishing)

Beihang BZK-005C/E

The BZK-005C/E is a Class III fixed-wing UCAV developed by Beihang University's UAV Institute (BUAA) and Harbin Aircraft Industry Group. Also known as the TYW-1, the BZK-005C/E is a strike-capable variant of the BZK-005, a high-altitude long-endurance surveillance drone developed in the mid-2000s. The BZK-005-series aircraft feature a single-engine, twin-boom design with mid-mounted wings and a low-mounted horizontal stabilizer. The BZK-005C/E are slightly heavier than their predecessors and are equipped with at least two wing-mounted hardpoints. BUAA and Harbin unveiled the strike-capable BZK-005C in a November 2017 ceremony and displayed the BZK-005E, an export-oriented version of the BZK-005C, at the 2018 Zhuhai Airshow. In November 2018, BUAA and Harbin released video footage of the BZK-005C/E conducting a weapons test with an undetermined type of air-to-ground munition. Although the unarmed BZK-005 is known to be in active service with China's PLA Navy and, potentially, the PLA Air Force, the BZK-005 does not appear to be in service with any domestic or foreign customers as of this writing.

The armed BZK-005 known as the TYW-1 was first shown at the Zhuhai Airshow 2018 and is seen here during weapons trials in a promotional video. (via SDF)

Specifications

Class	III
Status	A/B
Unveiled	2017
Operators	
Length	9.85m (32ft)
Width (wingspan)	18.0m (59ft)
Height	2.5m (8ft)
Maximum take-off weight	1,500kg (3,300lb)
Speed (max)	200km/h (108kts)
Range	
Endurance	40 hours
Flight ceiling	7,500m (25,000ft)
Payload	370kg (814lb)
Armament	

Fisher Jr., R. D., 'China's Beihang Unmanned Aircraft System Technology unveils TYW-1 strike-capable UAV', *Janes*, 16 November 2017, http://www.janes.com/article/75700/china-s-beihang-unmanned-aircraft-system-technology-unveils-tyw-1-strike-capable-uav

East Pendulum (@HenriKenhmann), 'Le drone BZK-005C…a mené récemment un exercice de tirs réels avec succès', Twitter, 3 December 2018, https://twitter.com/HenriKenhmann/status/1069550356349112321

CASC Caihong-3 (CH-3)

The Caihong-3, or CH-3, is a Class III fixed-wing UCAV developed by the China Aerospace Science and Technology Corporation (CASC). The CH-3 features swept-wing configuration with tricycle landing gear, two nose-mounted rectangular canards, and one hardpoint under each wing. CASC unveiled a full-scale mock-up of the CH-3 at the 2008 Zhuhai Airshow. In a video presentation at the 2010 Airshow, CASC described the CH-3 as a strike-capable platform compatible with the 45kg (99lb) AR-1 anti-tank missile and the 25kg (55lb) FT-10 guided bomb. The the armed version of the CH-3, is in military service with Algeria, Nigeria and Pakistan, though Islamabad operates a licensed variant of the CH-3 called the Burraq. Other international customers may include Myanmar, Sudan, and Turkmenistan, although it is not clear if the CH-3 is in service in an armed capacity. Chinese commercial mining interests have also used the CH-3 for surveying and mapping missions in Liberia and Zambia.

Allegedly a Pakistani Burraq, which is a licensed variant of the armed CH-3A, fitted with two AR-1 missiles. (dskaogu)

Specifications

Class	III
Status	A/A
Unveiled	2008
Operators	Algeria, Nigeria, Pakistan
Length	
Width (wingspan)	8.0m (26ft)
Height	
Maximum take-off weight	640kg (1,408lb)
Speed (maximum)	200km/h (108kts)
Range	200km (124 miles)
Endurance	12 hours
Flight ceiling	5,000m (16,400ft)
Payload	80kg (176lb)
Armament	AR-1, FT-5

MYCHINA, 'China new 'Rainbow' series UAV show in zhuhai/Zhuhai Air Show Rainbow series UAV', YouTube, 24 November 2010, https://www.youtube.com/watch?v=Bqeo3Hn6-W8

POCOCK, C., 'Zhuhai Airshow Includes Debut of China's J-10 Fighter', *Aviation International Online*, 20 November 2008, https://www.ainonline.com/aviation-news/defence/2008-11-20/zhuhai-airshow-includes-debut-chinas-j-10-fighter

STEVENSON, B., 'DSA12: China marches towards exports', *Shephard News*, 17 April 2012, https://www.shephardmedia.com/news/uv-online/dsa12-china-marches-towards-exports/

YOUTUBE, https://www.youtube.com/watch?v=Bqeo3Hn6-W8

CASC Caihong-4B (CH-4)

The Caihong-4B, or CH-4B, is a Class III fixed-wing UCAV developed by the China Aerospace Science and Technology Corporation (CASC). The CH-4B features what is now a conventional design for a medium-altitude long-endurance UCAV, with mid-mounted wings, a V-tail, and bulbous nose cone. CASC unveiled a full-scale model of the CH-4B at the 2012 Zhuhai Airshow, where it was displayed with the 45kg (99lb) AR-1 anti-tank missile and the 65kg (143lb) FT-5 guided bomb. The CH-4 has up to two hardpoints under each wing. In a video published by Chinese state television in 2015, the CH-4 appears equipped with one 50kg (110lb) CS/BBE2 satellite-guided bomb and another, unspecified 50kg bomb. In another video published by Chinese state television in November 2016, the CH-4 is shown conducting weapons trials with several missiles, including the AR-1. At the Zhuhai 2016 defence exhibition, CASC has displayed the CH-4 alongside the 16kg (35lb) TL-2, 85kg (187lb) TL-1, and 20kg (44lb) AR-2 air-to-surface guided missiles, which are designed specifically for use by UCAVs. The CH-4 is in military service with Algeria, Indonesia, Nigeria, Saudi Arabia, and Pakistan. Saudi Arabia has deployed the CH-4 to support operations in Yemen against the Houthi group. At least four CH-4s have been downed or crashed in Yemen. Other international customers include Jordan and Iraq, neither of which appear to continue to operate the CH-4.

An armed CH-4 was on display at Zhuhai in 2018 carrying an unknown sensor pod is 5 fitted with AR-1B and AR-2 missiles. In 2021 it was confirmed that the PLA Army Aviation was operating a similar variant. (Harpia Publishing)

Specifications

Class	III
Status	A/A
Unveiled	2012
Operators	Algeria, Indonesia, Nigeria, Pakistan, Saudi Arabia
Length	8.5m (28ft)
Width (wingspan)	18.0m (59ft)
Height	
Maximum take-off weight	1,330kg (2,900lb)
Speed (maximum)	180km/h (97.2kts)
Range	
Endurance	40 hours
Flight ceiling	
Payload	345kg (760lb)
Armament	AR-1, AR-2

ARTHUR, G., 'PREMIUM: Pakistan is confirmed as CH-4 UAV importer', *Shephard Media*, 5 February 2021, https://www.shephardmedia.com/news/uv-online/premium-pakistan-confirmed-ch-4-uav-importer/?utm_source=twitter&utm_medium=auto_post&utm_campaign=news_promo

CTGN, 'Watch: Exclusive strike video of Rainbow-4, China's home-developed UAV', YouTube, 3 November 2016, https://www.youtube.com/watch?v=MIHewBp-FRQ

DEFENCE FORUM INDIA, 'Airshow China 2012', *Defence Forum India* (blog), 8 November 2012, https://defenceforumindia.com/threads/airshow-china-2012.43639/

DRONEWARSUK, 'Drone Crash Database', *DroneWarsUK*, updated 19 January 2021, https://dronewars.net/drone-crash-database/

STEVENSON, B., 'DSA12: China marches towards exports', *Shephard News*, 17 April 2012, https://www.shephardmedia.com/news/uv-online/dsa12-china-marches-towards-exports/

ST (@aviation07101), 25 October 2020, https://twitter.com/aviation07101/status/1320454297910448129

'格调高：彩虹4挂50KG级小直径炸弹秀精准攻击_高清图集', *Mil.Sina*, 17 December 2015, http://slide.mil.news.sina.com.cn/k/slide_8_38692_39641.html

A CH-5 armed with AR-1B and AR-2 missiles on display at the Zhuhai Airshow in 2018. In December 2020 it was revealed that a UAV brigade in the Eastern Theater Command was using the type. (Harpia Publishing)

CASC Caihong-5 (CH-5)

The Caihong-5, or CH-5, is a Class III fixed-wing UCAV developed by the China Aerospace Science and Technology Corporation (CASC). Like its predecessor, the CH-4, the CH-5 features a standard design for medium-altitude long-endurance UCAVs with mid-mounted wings, a V-tail, a turbo-charged piston engine. The CH-5 is slightly larger than the CH-4 and, according to CASC, has an external payload capacity that around three times that of its predecessor. CASC revealed the CH-5 in a programme on Chinese state television in 2015, in which it announced that an initial CH-5 prototype had conducted its maiden flight. CASC exhibited the CH-5 to an international audience at the 2016 Zhuhai Air Show. In 2017, a revised and improved version of the CH-5 conducted its maiden flight. The CH-5 appears to have a similar compatible loadout to the CH-4. At Zhuhai in 2016, the CH-5 was displayed with the AR-1 and AR-2 air-to-ground missiles and the FT-9 guided bomb. In 2017, CASC announced that it has integrated a new, unspecified type of 80kg (176lb) precision guided munition on the CH-5. In photos published to social media in 2018, the CH-5 appeared equipped with an AR-1 missile and a larger, unspecified air-to-ground bomb.

Specifications (estimated)

Class	III
Status	B/B
Unveiled	2015
Operators	
Length	11.0m (36ft)
Width (wingspan)	21.0m (39ft)
Height	
Maximum take-off weight	3,000kg (6,500lb)
Speed (maximum)	300km/h (162kts)
Range (LOS)	250km (155 miles)
(BLOS)	6,500km (4,039 miles)
Endurance	30 hours
Flight ceiling	7,000m (22,966ft)
Payload	1,200kg (2,640lb)
Armament	

ARTHUR, G., 'China unveils 'game-changing' CH-5', *Shephard News*, 1 September 2015, https://www.shephardmedia.com/news/uv-online/china-unveils-game-changing-ch-5/

EAST PENDULUM (@HenriKenhmann), 'Le drone CH-5, conçu par l'Institut CAAA filiale du groupe d'aérospatiale chinois CASC', 1 May 2018, https://twitter.com/HenriKenhmann/status/991338401986101249

GADY, F-S, 'New Variant of China's CH-5 Combat Drone Boasts Extended Endurance and Range', *The Diplomat*, 5 April 2018, https://thediplomat.com/2018/04/new-variant-of-chinas-ch-5-combat-drone-boasts-extended-endurance-and-range/

KENHMANN, H., 'Le drone CH-5 teste deux nouvelles armes guidées', East Pendulum (blog), 27 September 2017, http://www.eastpendulum.com/drone-ch-5-teste-deux-nouvelles-armes-guidees

CASC Caihong-6 (CH-6)

The Caihong-6, or CH-6, is a conceptual Class III fixed-wing UCAV developed by the China Aerospace Science and Technology Corporation (CASC). CASC revealed a subscale model of the CH-6 in a post to social media in April 2021 and displayed a full-scale mockup of the aircraft at the Zhuhai Airshow in 2021. The CH-6's features mid-mounted, swept-back wings with winglets, retractable landing gear and a T-tail. Unlike the CASC CH-4 and CH-5, both of which feature propeller engines, the CH-6 employs two rear-mounted turbofans, enabling it to — at least on paper — reach speeds of up to 800km/h (432kts), more than twice that of the CH-5. The company-provided specifications for the CH-6 indicate that it will be a much heavier aircraft than its predecessors and capable of carrying up to 2,000kg (4,400lb) of payload. At Zhuhai in 2021, CASC displayed the CH-6 with the AR-1, AR-1B and AR-2 missiles. In what could be a first for UCAVs, the CH-6 was also exhibited alongside an air-dropped autonomous underwater vehicle (AUV) produced by the China Academy of Aerospace Aerodynamics, a CASC subsidiary. According to a CASC spokesperson, the CH-6's maiden flight is expected to occur in October 2022.

Specifications (strike configuration)

Class	III
Status	C/C
Unveiled	2021
Operators	
Length	15.0m (49ft)
Width (wingspan)	20.5m (67ft)
Height	5.0m (16ft)
Maximum take-off weight	7,800kg (17,160lb)
Speed (max)	800km/h (432kts)
Range	300km (186 miles)
Endurance	8 hours
Flight ceiling	12,000m (40,000ft)
Payload	2,000kg (4,400lb)
Armament	(AR-1, AR-1B, AR-2)

An armed CH-6 seen for the first time at Zhuhai 2021 and displayed alongside AR-1, AR-1B, and AR-2 missiles. (Taihushage)

Wong, K., 'Airshow China 2021: CASC unveils airdrop AUV concept', Janes, 30 September 2021,
https://www.janes.com/defence-news/news-detail/airshow-china-2021-casc-unveils-airdrop-auv-concept

CASC Caihong-7 (CH-7)

The Caihong-7 (彩虹-7, or CH-7) is a concept Class III fixed-wing UCAV developed by the 11th Research Institute of the China Aerospace Science and Technology Corporation (CASC). The CH-7 features a swept-back flying-wing design, dorsally mounted turbofan engine, and internal weapons and payload bays. CASC unveiled a full-scale mock-up of the CH-7 in 2018 at the Zhuhai Airshow, describing it as a high-altitude, long-endurance stealth combat UAV. At the time, it was scheduled to begin flight tests in 2019 and production by 2022, though it is not clear whether it has indeed flown. According to media reports, the CH-7 could be sold for export.

A mock-up on display at the Zhuhai Airshow in 2018 showing the CH-7 from above. Since then, nothing more has been revealed about its progress. (via SDF)

Specifications

Class	III
Status	C/C
Unveiled	2018
Operators	
Length	18.0m (33 ft)
Width (wingspan)	22.0m (72ft)
Height	
Maximum take-off weight	13,000kg (28,600lb)
Speed (cruise)	617km/h (333.2kts)
(maximum)	926km/h (500kts)
Range (operational)	
Endurance	
Flight ceiling	13,000m (42,000ft)
Payload	
Armament	

'China unveils stealth combat drone in development', *Associated Press*, 7 November 2018, https://www.cbc.ca/news/technology/china-combat-drone-1.4841792

CHUANREN, C., 'China Shows Off UAV Capabilities at Zhuhai', *Aviation International Online*, 9 November 2018, https://www.ainonline.com/aviation-news/defence/2018-11-09/china-shows-uav-capabilities-zhuhai

CASC CH-92A

The CH-92 is a Class II fixed-wing UCAV developed by the China Aerospace Science and Technology Corporation (CASC). The CH-92A features a twin-boom design with high-mounted wings and a mid-mounted tailplane. It has two external hardpoints, which can be equipped with the FT-8C or FT-8D air-to-ground missiles. It was displayed at the 2018 Zhuhai Air Show with both missile types. In August 2020, the Serbian military took delivery of six CH-92As and 18 FT-8C air-to-ground missiles. In October 2020, Serbia conducted a live-fire test of the CH-92A, also known as D-80 in local service.

Specifications

Class	II
Status	A/A
Unveiled	
Operators	Serbia
Length	4.3m (14ft)
Width (wingspan)	9.0m (29.ft)
Height	
Maximum take-off weight	300kg (660lb)
Speed (maximum)	190km/h (102.6kts)
Range (operational)	250km (155 miles)
Endurance	8 hours
Flight ceiling	5,950m (19,500ft)
Payload	75kg (165lb)
Armament	FT-8C

MINISTRY OF DEFENCE OF SERBIA, 'Minister Vulin at the first CH-92A unmanned aerial vehicle live-fire testing', news release, 21 October 2020, http://www.mod.gov.rs/eng/16641/ministar-vulin-na-prvom-bojevom-gadjanju-iz-bespilotne-letelice-ch-92a-16641

ROBLIN, S., 'Missile-Armed Chinese Drones Arrive in Europe as Serbia Seeks Airpower Edge', *Forbes*, 9 July 2020, https://www.forbes.com/sites/sebastienroblin/2020/07/09/missile-armed-chinese-drones-arrive-in-europe-for-serbian-military/?sh=eaf2c0479d24

ZOKA (@200_zoka), 'Real CH-92A. Photo by Mikhail Zherdev from Zhuhai-2018', Twitter, 28 November 2018, https://twitter.com/200_zoka/status/1200024225588940802/photo/1

A Serbian CH-92A, armed with FT-8C misiles, during an official presentation. (Dimitrije Ostojic)

CASC CH-901

The CH-901 is a lightweight fixed-wing loitering munition developed by the China Aerospace Science and Technology Corporation (CASC). The CH-901 is tube-launched and features missile-like fuselage with four foldable fins that expand upon launch. CASC exhibited the CH-901 at the 2016 Defence Services Asia exhibition, though it may have been revealed internationally in 2013 or earlier. At the 2018 Zhuhai Airshow, CASC displayed a launch vehicle with a capacity for 12 CH-901s. A video published in 2020 by CAEIT, a subsidiary of the state-owned firm CETC, demonstrated a launch vehicle with 48 cells for loitering munitions, potentially the CH-901 or another, similar aircraft.

Specifications

Class	Loitering munition
Status	A
Unveiled	2013
Operators	China
Length	1.2m (3.9ft)
Width (wingspan)	1.5m (4.9ft)
Height	
Maximum take-off weight	8.0kg (18lb)
Speed (maximum)	150km/h (81kts)
Range (operational)	15km (9 miles)
Endurance	1 hour
Flight ceiling	1,500m (4,920ft)
Payload	
Armament	Loitering munitions

Two CH-901 loitering munitions and a vehicle-mounted canister launch system displayed at the Military Museum in Beijing in 2017. (via Popular Science)

Armada, 'Compendium Drones', supplement to *armada* Issue 3/2014, Volume 38, June-July 2014
https://issuu.com/vishmeh/docs/armada_-_june_july_2014_drones_comp

East Pendulum (@HenriKenhmann), '| #AirshowChina 2018 | Le drone de reconnaissance tactique CH-901 peut être lancé maintenant en plusieur …',
Twitter, 7 November 2018, https://twitter.com/HenriKenhmann/status/1060208332663742465

Hambling, D., 'China's Mini-Drone Packs a Heavyweight Punch', *Popular Mechanics*, 6 May 2016,
https://www.popularmechanics.com/military/a20722/china-mini-drone/

Hambling, D., 'China Releases Video Of New Barrage Swarm Drone Launcher', *Forbes*, 14 October 2020,
https://www.forbes.com/sites/davidhambling/2020/10/14/china-releases-video-of-new-barrage-swarm-drone-launcher/?sh=dd8e40f2ad7f

CASC V750/QY-1

The V750 is a Class III rotary-wing UCAV developed by the China Aerospace Science and Technology Corporation (CASC). It features a single main rotor and a tail rotor. It is based on the manned, US-made Brantly B-2 light helicopter. CASC presented an unarmed mock-up of the V750 at the Zhuhai air show in November 2010. According to Shephard News, the V750 conducted its maiden flight in May 2011. Photos published to social media in June 2016 appeared to show an armed variant of the V750 conducting missile tests. CASC presented the QY-1, an armed variant of the V750, at the Zhuhai air show in November 2016, where it was fitted with two 25kg (55lb) FT-10 air-to-ground bomb on two outrigger pylons.

Specifications

Class	III
Status	B/B
Unveiled	2010
Operators	
Length	6.43m (21ft)
Width	
Height	2.11m (6.9ft)
Maximum take-off weight	758kg (1,670lb)
Speed (maximum)	161km/h (86.9kts)
Range (LOS)	150km (93 miles)
Endurance	
Flight ceiling	
Payload	120kg (265lb)
Armament	FT-10

A V750 during missile tests. (SDF)

Lin, J. and Singer, P. W., 'China's armed robot helicopter takes flight', *Popular Science*, 11 July 2016,
https://www.popsci.com/chinas-armed-robot-helicopter-takes-flight/

Minnick, W., 'Zhuhai 2016: Chinese-American UAV helo weaponised', *Shephard News*, 3 November 2016,
https://www.shephardmedia.com/news/uv-online/zhuhai-2016-chinese-american-uav-helo-weaponised/

Waldron, G., 'Zhuhai10: Pictures: China reveals armed UAV designs', *FlightGlobal*, 25 November 2010,
https://www.flightglobal.com/zhuhai10-pictures-china-reveals-armed-uav-designs/97102.article

CASIC Tian Ying

The Tian Ying (天鹰) is a demonstrator Class III fixed-wing UAV developed by the 302nd Institute of the China Aerospace Science Industry Corporation (CASIC). Like the CH-7, the Tian Ying has a tailless flying-wing design, though it is only half the size of the CH-7, with an estimated 10m (33ft) wingspan. CASIC published blurred photos of an aircraft thought to be the Tian Ying in early 2018, reporting that it had conducted its first flight tests in late 2017. It was officially revealed at the 2018 Zhuhai Airshow. In January 2019, CASIC published a video of the Tian Ying conducting flight tests at Baotou 2 Airport in Inner Mongolia province. The Tian Ying is believed to be designed to engage in both armed and unarmed missions, though as of this writing it is not clear if it is has carried out any weapons trials.

Specifications

Class	III
Status	B/C
Unveiled	2018
Operators	
Length	
Width (wingspan)	10.0m (33ft)
Height	
Maximum take-off weight	
Speed (maximum)	675km/h (364.5kts)
Range (operational)	
Endurance	15 hours
Flight ceiling	12,800m (42,000ft)
Payload	
Armament	

The appearance of the Tian Ying at the Zhuhai show in 2018 was a surprise. As with other Chinese flying wing UAV/UCAVs, little is known about its current status. (via SDF)

East Pendulum (@HenriKenhmann), 'Extrait d'une vidéo de CASIC Institut 302, qui montre pour la première fois le décollage et l'atterrissage du drone à aile volant 'Tian Ying' (天鹰)', Twitter, 1 January 2019, https://twitter.com/HenriKenhmann/status/1080140472868167681

Kenhmann, H., 'CASIC fait voler son nouveau drone furtif à aile volante', East Pendulum, 12 February 2018, http://www.eastpendulum.com/casic-fait-voler-son-nouveau-drone-furtif-a-aile-volante

Trevithick, J. and Rogoway, T., 'China's Biggest Airshow Offers More Evidence Of Beijing's Stealth Drone Focus', *The Drive*, 2 November 2018, http://www.thedrive.com/the-war-zone/24645/chinas-biggest-airshow-offers-more-evidence-of-beijings-stealth-drone-focus

CASIC WJ-600A/D

The WJ-600A/D is a Class III fixed-wing UCAV developed by China Aerospace Science and Industry Corporation (CASIC). The WJ-600A/D is a strike-capable variant of the WJ-600. The WJ-600-series feature a slender, missile-like fuselage with high-mounted wings and a single turbofan jet engine. The WJ-600A/D features one hardpoint mounted under each wing. It is vehicle-launched and recovered by parachute. CASIC has advertised the WJ-600A/D as a high-speed UCAV designed to evade radar detection. CASIC unveiled the WJ-600A/D at the 2010 Zhuhai Airshow, where it was displayed with KD-2 and TB-1 missiles and a ZD-1 bomb, though it is unclear whether any of these munitions have been tested with the WJ-600A/D. In a 2016 military parade, Turkmenistan displayed multiple WJ-600A/Ds, each of which appeared equipped with the CM-502KG, a miniaturized version of the CM-501 air-to-ground guided missile. It is unclear whether the WJ-600A/D remains in active service with Turkmenistan or any other country.

Specifications

Class	III
Status	A/A
Unveiled	2010
Operators	China, Turkmenistan
Length	
Width (wingspan)	
Height	
Maximum take-off weight	
Speed (max)	700km/h (378kts)
Range	
Endurance	3.2 hours
Flight ceiling	8,000m (26,240ft)
Payload	130kg (286lb)
Armament	CM-502KG

Defence Blog, 'Turkmenistan took delivery Chinese high speed WJ 600 unmanned aerial vehicle', YouTube, 28 October 2016, https://www.youtube.com/watch?v=kIcQVXsMmms

Waldron, G., 'Zhuhai 10: Pictures: China reveals armed UAV designs', *FlightGlobal*, 25 November 2010, https://www.flightglobal.com/zhuhai10-pictures-china-reveals-armed-uav-designs/97102.article

A rare image of a WJ-600 from an advertising placard showing the UAV during both launch and landing. (via CDF)

GAIC Yaoying-2/Air Sniper

The Yaoying-2 (鹞鹰-II) is a demonstrator Class III fixed-wing UAV developed by GAIC, a subsidiary of AVIC. Also known as the 'Air Sniper', the Yaoying-2 is a strike-capable system with a 16-hour endurance and 200km (124-mile) operational radius. It is the successor to the Yaoying-1, which was unveiled in 2011. AVIC unveiled a subscale model of the Yaoying-2 at the 2014 Zhuhai Air Show. The Yaoying-2 conducted its maiden flight at Anshun Airport in July 2018. In photographs of these tests, the Yaoying-2 was pictured with two Blue Arrow-7 air-to-ground guided missiles, one equipped to a hardpoint under each wing. AVIC also displayed the Yaoying-2 at the 2019 Guiyang Industrial Products Fair equipped with two Blue Arrow-7 missiles. According to Chinese media reports, AVIC is hoping to market the aircraft to foreign customers. AVIC displayed a subscale model of the planned Yaoying-3, a twin-engine variant with a flying wing design and internal weapons bay, at the 2012 Zhuhai Air Show, although AVIC has not announced any progress on the development of this variant.

Specifications

Class	III
Status	B/B
Unveiled	2014
Operators	
Length	7.4m (24ft)
Width (wingspan)	14.4m (47ft)
Height	2.77m (9ft)
Maximum take-off weight	1,250kg (2,750lb)
Speed (maximum)	150km/h (81kts)
Range	
Endurance	16 hours
Flight ceiling	
Payload	250kg (550lb)
Armament	Blue Arrow 7

China News Service, 'On July 26, the first Guiyang Industrial Products Fair was held at the Guiyang International Convention and Exhibition Centre', Facebook, 26 July 2019, https://www.facebook.com/ChinaNewsService/posts/2558562270844988/

Shen, A., 'Chinese firm eyes overseas market as strike drone Yaoying-2 makes maiden flight', *South China Morning Post*, 5 July 2018, https://www.scmp.com/news/china/diplomacy-defence/article/2153963/chinese-firm-eyes-overseas-market-strike-drone-yaoying

The Yaoying-2 on display at Zhuhai in 2018. (Harpia Publishing)

Harwar Zhanfu H16-V12

The Zhanfu H16-V12, also known as Tomahawk or simply H16, is a Class I rotary-wing UCAV developed by Harwar, a Shenzhen-based UAV manufacturer. The H16, a heavy lift hexacopter, features a coaxial multirotor design. It can be equipped with various payloads for public safety and military missions. Harwar unveiled the H16 at the 2018 Shenzhen International Drone Exhibition. Harwar has displayed the H16 at multiple domestic and international defence exhibitions alongside a light machinegun, miniature laser-guided munitions, and grenade launchers. According to a January 2020 report in *Janes*, the H16 may already be in service with the People's Liberation Army.

Specifications

Class	I
Status	A/A
Unveiled	2018
Operators	China
Length	1.53m (5ft)
Width	1.75m (5.8ft)
Height	0.83m (3ft)
Maximum take-off weight	
Speed (maximum)	65km/h (35.1kts)
Range (operational)	14.4km (9 miles)
Endurance	1 hour
Flight ceiling	5,800m (19,000ft)
Payload	25kg (55lb)
Armament	

The Harwar H16-V12 is known to be in use by both armed police forces, but also with PLA Army units. (via SDF)

DOMINGUEZ, G., 'Harwar's Zhanfu H16-V12 UAV possibly in service with Chinese military', *Janes*, 24 January 2020,
https://www.janes.com/defence-news/news-detail/harwars-zhanfu-h16-v12-uav-possibly-in-service-with-chinese-military

HiWING WJ-700 Lieying

The WJ-700 Lieying (Falcon) is a demonstrator Class III fixed-wing UCAV developed by HiWING, a subsidiary of the China Aerospace Science and Industry Corporation. The WJ-700 features a conventional MALE UAV design with mid-mounted wings and a V-tail. It is equipped with a turbojet engine. CASIC unveiled a full-scale mock-up of the WJ-700 at the 2018 Zhuhai Airshow, where it was displayed with the Blue Arrow-21 anti-tank guided missile, the C-705KD anti-ship missile, and the 80kg (176lb) CM-102 anti-radiation missile. The mock-up of the WJ-700 featured two hardpoints under each wing. In January 2021, CASIC announced that the WJ-700 had completed its maiden flight.

Specifications

Class	III
Status	B/C
Unveiled	2018
Operators	
Length	10.0m (33ft)
Width (wingspan)	
Height	
Maximum take-off weight	3,500kg (7,700lb)
Speed (maximum)	700km/h (378kts)
Range (operational)	
Endurance	20 hours
Flight ceiling	12,000m (39,460ft)
Payload	
Armament	(Blue Arrow-21), (C-705KD), (CM-102)

The WJ-700 fitted with CM-502KG missiles and displayed alongside the AG-500M, C-705KD and CM-501X missiles at the 2021 Zhuhai Airshow. (Iron Eagle)

Waldron, G., 'PICTURE: CASIC shows off new long-range strike UAV', *FlightGlobal*, 5 November 2018,
https://www.flightglobal.com/military-uavs/picture-casic-shows-off-new-long-range-strike-uav/130134.article

Wong, K., 'China's WJ-700 Falcon armed reconnaissance UAV makes maiden flight', *Janes*, 12 January 2021,
https://www.janes.com/defence-news/news-detail/chinas-wj-700-falcon-armed-reconnaissance-uav-makes-maiden-flight

NORINCO CR-500 Golden Eagle

The CR-500 Golden Eagle is a Class II rotary-wing UCAV developed by the China North Industries Group Corporation Limited (NORINCO). It features coaxial rotors and a stubby fuselage. NORINCO unveiled the CR-500 at the 2016 Zhuhai air show, where it was fitted with two Blue Arrow-9 air-to-ground missiles. At the 2018 Zhuhai air show, NORINCO displayed the CR-500 fitted with eight missiles, possibly the Red Arrow-10 (HJ-10) anti-tank missile. At the UMEX defence exhibition in February 2020, NORINCO displayed the CR-500 alongside several Blue Arrow-7 missiles. In November 2020, NORINCO announced on the Chinese social media site WeChat that the company had completed preparing the CR-500 for delivery to an undisclosed customer. In February 2021, Breaking Defense reported that the United Arab Emirates was planning on purchasing '10 to 15 Golden Eagle CR500 helo drones fitted with Red Arrow 12 missiles' for a total of USD 9 million, according to an unnamed official, though this has yet to be confirmed.

Specifications

Class	II
Status	B/B
Unveiled	2016
Operators	
Length	
Width	
Height	
Maximum take-off weight	470kg (1,034lb)
Speed (maximum)	140km/h (75.6kts)
Range	200km (124 miles)
Endurance	5 hours
Flight ceiling	
Payload	160kg (352lb)
Armament	HJ-12, (HJ-10), (Blue Arrow-7), (Blue Arrow-9)

The CR-500 employs an unusual coaxial rotor design. Even if reported in late 2020 to be ready for delivery, the customer is not yet disclosed. This fully armed CR-500 was seen at Zhuhai in 2018. (Harpia Publishing)

Dafeng cao (@dafengcao), 'Golden Eagle-CR500 recon & strike unmanned helicopter', Twitter, 30 October 2016,
https://twitter.com/dafengcao/status/792619843966996480

DefenceWebTV, 'UMEX 2020 International Unmanned Defence Systems and Training Exhibition Abu Dhabi UAE Day 3', YouTube, 25 February 2020,
https://www.youtube.com/watch?v=rSFbKL6AEG4

Mezher, C., 'China, UAE Forge New Deals; Joint R&D Plan & New Drone Sales', *Breaking Defense*, 24 February 2021,
https://breakingdefense.com/2021/02/china-uae-forge-new-deals-joint-rd-plan-new-drone-sales/

Mizokami, K., 'All the New Tech from China's Big Air Show', *Popular Mechanics*, 6 November 2018,
https://www.popularmechanics.com/military/aviation/a24680684/all-the-new-tech-from-chinas-big-airshow/

Rupprecht, A. and Dominguez, G., 'Norinco's CR500 VTOL UAV cleared for delivery to undisclosed customer', *Janes*, 30 November 2020,
https://www.janes.com/defence-news/news-detail/norincos-cr500-vtol-uav-cleared-for-delivery-to-undisclosed-customer

NORINCO MR-40 and MR-150

The MR-40 and MR-150 are two rotary-wing UCAVs developed by the China North Industries Group Corporation Limited (NORINCO). Both are multirotor drones; the MR-40 features four rotors and the MR-150 sports six. NORINCO unveiled the MR-40 and MR-150 at the Zhuhai Airshow in November 2018. NORINCO displayed an MR-40 at Zhuhai 2018 fitted with two lightweight air-to-ground missiles and an MR-150 equipped with two 11kg (24lb) BBE2 miniature bombs. A second MR-150 was shown equipped with two missile tubes that could reportedly accommodate the 20kg (44lb) Blue Arrow-5 air-to-ground guided missile. Both aircraft could also be equipped with LG5A 40mm grenade launchers. In February 2021, Breaking Defense reported that the United Arab Emirates was planning on purchasing 20 MR-40s equipped with BBE2 bombs for a total of USD 7 million. It is not clear, however, if either aircraft has conducted flight or weapons trials.

Specifications

	MR-40
Class	I
Status	C/C
Unveiled	2018
Operators	
Length	
Width	
Height	
Maximum take-off weight	
Speed (maximum)	
Range	5.0km (3 miles)
Endurance	0.5 hours
Flight ceiling	
Payload	
Armament	BBE2

Aiming for a similar role as the Harwar H16-V12, little is known about either the MR-40 or the MR-150. (via SDF)

Tengden Featherless Arrow series

Sichuan Tengden Technology Co. has produced a family of prototype Class II rotary-wing UAVs known as the Featherless Arrow series (没羽箭). Tengden unveiled the first elements of the Featherless Arrow series at the 14th China-ASEAN Expo in 2017, displaying full-size mockups of the HA-001 and HB-001 aircraft. The HA-001 and HB-001 had a reported maximum take-off weight of 450kg (992lb) and 280kg (617lb), respectively, and were both displayed equipped with the FT-8D lightweight air-to-ground missile on two outrigger pylons. The HA-001 conducted its first flight in a September 2020 test at Kangding Airport in Sichuan province. In flight tests at Kangding Airport in December 2020 and May 2021, Tengoen conducted revealed two additional Featherless Arrow aircraft, the HA-002 and TH-145, both of which closely resemble the HA-001 in terms of technical specifications and physical characteristics and likely represent improved or modified versions of the original. According to Tengden, the HA-002 has a maximum take-off weight of 550kg (1,213lb), slightly more than the HA-001 presented in 2017. Although the HA-001 and HB-001 mockups were displayed in 2017 with the FT-8D missile, it does not appear as though the Featherless Arrow series aircraft have conducted weapons trials as of this writing. The Featherless Arrow aircraft remain in development and do not appear to be in active military service.

Specifications

	HA-002
Class	II
Status	B/C
Unveiled	2017
Operators	
Length	7.9m (26ft)
Width (wingspan)	6.4m (21ft)
Height	
Maximum take-off weight	550kg (1,210lbs)
Speed (max)	
Range	900km (558mi)
Endurance	8 hours
Flight ceiling	6,500m (21,320ft)
Payload	120kg (264kg)
Armament	(FT-8D)

The Tengden HB-001 is a rotorcraft UCAV similar in role and configuration to the AV500W. In common with its competitor, little is known about its current status. (Harpia Publishing)

Tengden Technology TA-001

The TA-001 is a demonstrator Class III fixed-wing UCAV developed by Tengden Technology. The TA-001 is a smaller version of the TB-001. The TA-001, which is also known as the Flying Eagle (的扑天雕), has a twin-boom design with high-mounted wings and a high-mounted tailplane. Unlike the TB-001, the TA-001 has a single engine in the pusher configuration. It was unveiled at the 14th China-ASEAN Expo in 2017 in Nanning, where it was displayed with two air-to-ground munitions – possibly the FT-8B or FT-8C – each of which were attached to a hardpoint under each wing. The TA-001 conducted its first flight in February 2018. It is not clear whether it is in active service with any military.

Specifications

Class	III
Status	B/C
Unveiled	2017
Operators	
Length	7.8m (26ft)
Width (wingspan)	14.7m (48ft)
Height	2.7m (9ft)
Maximum take-off weight	1,200kg (2,640lb)
Speed (maximum)	
Range	
Endurance	24 hours
Flight ceiling	7,500m (24,600ft)
Payload	300kg (660lb)
Armament	(FT-8B, FT-8C)

In contrast to the larger TB-001, the TA-001 has so far attracted much less interest and its current status is unclear. (Tengden)

Tengden Technology TB-001/TW-328

The TB-001 is a Class III fixed-wing drone developed by Tengden Technology. Also known as the TW-328 or as the 'Twin Tailed Scorpion' (双尾蝎), the TB-001 initially featured a twin-engine, twin-boom design with high-mounted wings and a high-mounted vertical stabilizer. It was unveiled at the 14th China-ASEAN Expo in 2017 in Nanning. It conducted its first flight tests at Liangping Airport in Chongqing province in 2017. SF Express, the Chinese package delivery firm, has conducted field tests of a TB-001 modified for cargo operations in Yunnan province. In January 2020, Tengden announced that it had conducted tests of a three-engine variant of the TB-001. This unnamed variant has a maximum take-off weight of 3,200kg (7,055lb), 400kg (882lb) more than the two-engine TB-001, and a flight ceiling of 9,500m (31,168ft), 1,500m (4,921ft) more than its predecessor. Tengden has described the TB-001 as a combat UAV. The TB-001 was displayed at ASEAN 2017 alongside the FT-7 glide bomb, FT-8D guided missile, and FT-9 and FT-10 guided bombs. It was displayed at Zhuhai 2018 with the C-701/K anti-ship missile, and CM-502 air-to-surface missile, and the FT-9 guided bomb. Photos published to the online forum Airliners.net in October 2019 showed a parked TB-001 equipped with eight air-to-ground munitions of varying size, potentially of the AR family. Indeed, Chinese media reports suggest that the TB-001 can be equipped with the 20kg (44lb) AR-2 and 80kg (176lb) AR-4 air-to-surface-missile, as well as the 100kg (220lb) AR-3 cruise missile. The TB-001 does not appear to be in active military service with any country as of this writing. Science Technology for Development and Industrial Investment, a Saudi Arabian company, has marketed a licensed variant of the TB-001 to customers in the Middle East, displaying a subscale model of the aircraft at the UMEX 2020 and IDEX 2021 defence exhibitions.

Specifications

Class	III
Status	B/B
Unveiled	2017
Operators	
Length	11.0m (36ft)
Width (wingspan)	20.0m (66ft)
Height	3.3m (11ft)
Maximum take-off weight	2,800kg (6,160lb)
Speed (maximum)	
Range (LOS)	280km (151.2kts)
(BLOS)	3,000km (1,865 miles)
Endurance	35 hours
Flight ceiling	8,000m (26,200ft)
Payload	
Armament	

The TB-001, or TW-328, seen here at the 2021 Zhuhai Airshow, is among the largest UAVs produced in China. (Camera)

Chuanren, C., 'New Chinese Company Shows Family of UAVs', *Aviation International Online*, 6 October 2017,
https://www.ainonline.com/aviation-news/defence/2017-10-06/new-chinese-company-shows-family-uavs

Lin, J. and Singer, P. W., 'China's new drone company is building a UAV with a 20-ton payload', *Popular Science*, 23 January 2018,
https://www.popsci.com/chinas-new-drone-company-has-big-plans

Rupprecht, A., 'Three-engined variant of China's Tengden TB001 UAV makes maiden flight', *Janes*, 21 January 2020,
https://www.janes.com/defence-news/news-detail/three-engined-variant-of-chinas-tengden-tb001-uav-makes-maiden-flight

Wong, K., 'Tengden readies production-ready TB001 armed reconnaissance UAV', *Janes*, 28 March 2019,
https://www.janes.com/defence-news/news-detail/tengden-readies-production-ready-tb001-armed-reconnaissance-uav

Ziyan Blowfish A2/3

The Blowfish A2 is a Class I rotary-wing UCAV developed by Ziyan Unmanned Aerial Vehicle Company, a Zhuhai-based firm. Ziyan published images on social media of the initial, unarmed Blowfish design conducting flight tests in September 2017. Ziyan initially described the Blowfish as a commercial VTOL UAV for transport, logistics, and emergency rescue. In September 2018, Ziyan displayed the Blowfish A1 at AAD 2018 defence exhibition in South Africa equipped with three mortar shells. Ziyan unveiled the Blowfish A2, which has a more modular payload bay than its predecessor, in 2019. Ziyan displayed the Blowfish A2 at the IDEX 2019 defence exhibition alongside mortar shells, a 40-millimeter grenade launcher, and a light machine gun. The Blowfish A3 features a few minor design differences to the A2 such as the addition of landing skids, but otherwise bears a strong resemblance to the A2. In November 2020, citing a local media report, *Janes* reported that the Tibet Military District of the People's Liberation Army Ground Force had ordered several Blowfish A2s.

Specifications

Class	I
Status	A/A
Unveiled	2017
Operators	China, Indonesia
Length	2.0m (6.6ft)
Width	
Height	
Maximum take-off weight	38kg (84lb)
Speed (maximum)	140km/h (75.6kts)
Range (operational)	
Endurance	1 hour
Flight ceiling	2,800m (9,184ft)
Payload	12kg (26lb)
Armament	40mm grenade, light machine gun, mortars

The Blowfish A3 is a smaller rotary UCAV and, although unconfirmed, it seems that it has been in PLA Army service since late 2020. (SDF)

Act (@zaylog), 'BLOWFISH 河豚 A2'という中国 ZIYAN（紫燕）UAV 社製のドローンです',Twitter, 24 April 2019, https://twitter.com/zaylog/status/1120998501251289088

Dominguez, G., 'PLAGF's Tibet Military District orders new VTOL UAVs, says report', *Janes*, 19 November 2020, https://www.janes.com/defence-news/news-detail/plagfs-tibet-military-district-orders-new-vtol-uavs-says-report

ZT Guide Fei Loong-1

The Fei Long-1 (飞龙-1) is a Class III fixed-wing UAV developed by Zhong Tian Guide Control Technology Company (ZT Guide). The Fei Long-1 features a V-tail and mid-mounted wings. ZT Guide unveiled the Fei Long-1 at the 2018 Zhuhai Airshow, describing it as a 'Large Payload Long Endurance Universal Unmanned Transportation Platform'. In January 2019, ZT Guide announced that it had conducted its maiden flight in a test at Pucheng Neifu Airport in Shaanxi province. ZT Guide displayed a subscale model of the Fei Long-1 at the 2020 UMEX defence exhibition in Abu Dhabi that was equipped with multiple munitions, suggesting that an armed, export-oriented variant of the Fei Long-1 may be under development. However, as of this writing it is unclear whether the Fei Long-1 has conducted weapons trials.

Specifications

Class	III
Status	B/C
Unveiled	2018
Operators	
Length	10.0m (33ft)
Width (wingspan)	20.0m (65.6ft)
Height	
Maximum take-off weight	3,200kg (7,040lb)
Speed (maximum)	320km/h (172.8kts)
Range (LOS)	250km (155 miles)
(BLOS)	2,000km (1,242 miles)
Endurance	45 hours
Flight ceiling	8,000m (26,240ft)
Payload (incl. fuel)	1,400kg (3,080lb)
Armament	

ZT Guide has several different UAVs under development and it seems the Fei Loong-1 has at least made its maiden flight. Otherwise its current status is unknown. (SDF)

BY78, 'Chinese UCAVs at UMEX in Abu Dhabi', Sino Defence Forum, 13 March 2020, https://www.sinodefenceforum.com/t/chinese-uav-ucav-development.3526/page-375

UMEX Abu Dhabi, https://umexabudhabi.ae/exhibit/2020-exhibitor-list/#203

Wong, K., 'China's FL-1 MALE UAV performs maiden flight', *Jane's International Defence Review*, 22 January 2019, https://www.janes.com/article/85869/china-s-fl-1-male-uav-performs-maiden-flight

European Union (France, Germany, Italy and Spain)

EURODRONE

The EuroDrone is a concept Class III fixed-wing UAV developed by a European consortium led by Airbus with participation from Dassault Aviation, and Leonardo. Also known as the EuroMALE or MALE RPAS, the EuroDrone is a joint project between Germany, France, Italy, and Spain. A full-scale mock-up of the EuroDrone was unveiled at the 2018 ILA Berlin air show. Its current design features low-mounted wings, a T-tail, and twin turboprop engines. Although the EuroDrone has been conceptualized as primarily serving an intelligence-gathering function, it is probable that some of its eventual users will decide to arm it should the project make it to completion. In a May 2019 event, OCCAR, the European arms procurement agency, displayed a subscale model of the EuroDrone that was fitted with munitions. In mid-2020, Aviation Week reported that Germany was planning on arming the EuroDrone with Paveway guided bombs and MBDA Brimstone air-to-ground guided missiles, though political opinion in Berlin over the future of armed drones remains divided. In late 2020, OCCAR concluded negotiations with Airbus over the final schedule and cost of the EuroDrone. Beginning in the late 2020s, Airbus is expected to deliver 20 EuroDrone systems to European militaries, each of which will comprise of three aircraft and associated equipment. The EuroDrone's maiden flight is scheduled for 2025.

Specifications

Class	III
Status	C/C
Unveiled	2018
Operators	
Length	16.0m (53ft)
Width (wingspan)	26.0m (85ft)
Height	6.0m (20ft)
Maximum take-off weight	10,000kg (22,000lb)
Speed (maximum)	500km/h (270kts)
Range	
Endurance	
Flight ceiling	13,700m (45,000ft)
Payload	2,300kg (5,060lb)
Armament	(AGM-114), (GBU-12 Paveway), (Brimstone)

AVIATION WEEK STAFF, 'Germany 'Plans Paveway, Brimstone for Euro-Drone', *Aviation Week*, 1 June 2020, https://aviationweek.com/defence-space/aircraft-propulsion/germany-plans-paveway-brimstone-eurodrone

DONALD, D., 'EuroDrone looks to development and production contract', *Aviation International Online*, 14 December 2020, https://www.ainonline.com/aviation-news/defence/2020-12-14/eurodrone-looks-development-and-production-contract/

OCCAR, 'European MALE RPAS Stage 2 Offer Submitted', news release, 29 May 2019, https://www.occar.int/european-male-rpas-stage-2-offer-submitted-29-may-2019/

POCOCK, C., 'Publicity Boost in Berlin for the Euro-MALE Drone', *Aviation International Online*, 26 April 2018, https://www.ainonline.com/aviation-news/defence/2018-04-26/publicity-boost-berlin-euro-male-drone/

WIEGOLD, T., 'Regierungskoalition macht Weg frei für Entwicklung der Eurodrohne', Augen geradeaus (blog), 3 February 2021, https://augengeradeaus.net/2021/02/regierungskoalition-macht-weg-frei-fuer-entwicklung-der-eurodrohne/

The Airbus European MALE Eurodrone mock-up at the ILA airshow in Berlin in 2018. (Airbus)

FRANCE

Safran Patroller

The Patroller is a Class III-A fixed-wing UAV developed by Safran Electronics and Defence. In 2016, the Direction générale de l'armement (DGA), the French military procurement agency, awarded Safran a USD 324 million contract for 14 Patroller aircraft and associated support equipment to replace the French Army's Sagem Sperwer. Although the original contract was for unarmed Patrollers for surveillance and reconnaissance, the French Army has since expressed an interest in arming the UAVs. In testimony before French legislators in 2019, Army chief Gen Thierry Burkhad said that the Army would pursue plans to arm the Patroller. In late 2019, the DGA awarded Safran a contract to integrate the Thales Aculeus 68 mm, a laser-guided induction rocket system, into the Patroller. The Patroller was also displayed at the 2019 Paris Air Show with the Aculeus rocket. However, neither the DGA nor Safran have confirmed that the Patroller has conducted a live fire test with the Aculeus rockets. The first unarmed Patrollers are expected to be delivered to the French Army in 2021.

Specifications

Class	III
Status	B/C
Unveiled	2009
Operators	France
Length	8.5m (28ft)
Width (wingspan)	18.0m (59ft)
Height	2.5m (8ft)
Maximum take-off weight	1,100kg (2,420lb)
Speed (maximum)	210km/h (113.4kts)
Range (LOS)	200km (124.3 miles)
(BLOS)	1,000km (620 miles)
Endurance	15 hours
Flight ceiling	4,500m (15,000ft)
Payload	200kg (440lb)
Armament	Aculeus 68mm

CABIROL, M., 'Le drone 'Patroller sera armé, l'armée de Terre le souhaite' (Chef d'état-major de l'armée de Terre)', *La Tribune*, 14 November 2019, https://www.latribune.fr/entreprises-finance/industrie/aeronautique-defence/le-drone-patroller-sera-arme-l-armee-de-terre-le-souhaite-chef-d-etat-major-de-l-armee-de-terre-833107.html

TAN, P., 'Sagem Patroller beats out Thales Watchkeeper in French Army drone pick', *Defense News*, 22 January 2016, https://www.defensenews.com/home/2016/01/22/sagem-patroller-beats-out-thales-watchkeeper-in-french-army-drone-pick/

VALPOLINI, P., 'Safran arms Patroller UAV', *Janes*, 30 June 2020, https://www.janes.com/defence-news/news-detail/safran-arms-patroller-uav

Safran Patroller. (Sagem)

GEORGIA

TAM T-31

The T-31 is a concept for a Class III fixed-wing UCAV developed by TAM Management. In January 2021, TAM Management revealed a computer-generated illustration of the planned T-31 UCAV that featured high-mounted wings, a tractor-propeller configuration, and a V-tail. According to TAM Management, the T-31 will have a 350kg (772lb) external payload capacity, which could accommodate additional fuel tanks or missiles. A January 2021 program on Georgian Business Media television including addition computer-generated drawings of the T-31, showing is designed to carry two missiles under each wing.

Specifications

Class	III
Status	C/C
Unveiled	2021
Operators	
Length	
Width (wingspan)	
Height	
Maximum take-off weight	
Speed (maximum)	
Range (LOS)	
Endurance	24 hours
Flight ceiling	6,000m (20,000ft)
Payload	350kg (770lb)
Armament	

A screen shot of Georgia's T-31 UCAV.

Bozinovski, I., 'Georgia developing indigenous surveillance and strike UAS', *Janes*, 12 February 2021, https://www.janes.com/defence-news/news-detail/georgia-developing-indigenous-surveillance-and-strike-uas/

RFERL, 'Georgia is going to produce unmanned aerial vehicles', YouTube, 2 February 2021, https://www.youtube.com/watch?v=Hin98XTCGvs&feature=emb_title/

INDIA

HAL Combat Air Teaming System 'Warrior'

The Combat Air Teaming System (CATS) Warrior is a conceptual Class III fixed-wing UCAV developed by Hindustan Aeronautics Limited (HAL). It is part of the CATS family of loyal wingman aircraft and associated systems. Other members of the CATS family include the ALFA, a small swarming UAV, and the Hunter, a weapons carrier that can be fitted on manned fighter aircraft. The CATS family will, in theory, eventually operate alongside the CATS Mothership for Air teaming eXploitation (MAX), a modified HAL Tejas Mk1A manned fighter that will be designed to control the loyal wingman and smaller UAVs. HAL unveiled the CATS family in February 2021 at the Aero India defence exhibition, where it displayed a full-scale mock-up of the Warrior, a loyal wingman UCAV that is the largest member of the CATS family. The Warrior features a low-observable design with a V-tail and a PTAE-7 turbojet engine. At the 2021 Aero India exhibition, the Warrior mock-up appeared equipped with the DRDO Smart Anti-Airfield Weapon (SAAW), a 120kg (265lb) guided glide bomb. It was displayed alongside the MBDA ASRAAM short-range air-to-air missile, known in India as the NGCCM. HAL is aiming to start flight testing the CATS family in the mid-2020s.

Specifications

Class	III
Status	C/C
Unveiled	2021
Operators	
Length	
Width (wingspan)	
Height	
Maximum take off weight	1,300kg (2,860lb)
Speed (maximum)	860 km/h (464.4kts)
Range	800km (496 miles)
Endurance	1.4 hours
Flight ceiling	
Payload	250kg (550lb)
Armament	

Combat Air Teaming System (CATS) Warrior. (Angad Singh)

CHANDRA, A., 'HAL unveils ambitious air-teaming system centred on Tejas', *FlightGlobal*, 3 February 2021,
https://www.flightglobal.com/defence/hal-unveils-ambitious-air-teaming-system-centred-on-tejas/142280.article

INDONESIA

Enrol Sistem Enrol Pilot

The Enrol Pilot is a fixed-wing loitering munition developed by PT Enrol Sistem Indonesia. The Enrol Pilot features a standard flying wing design with a rounded fuselage. It employs a single electric propeller in the pusher configuration and is hand-launched. Enrol Sistem, an electronics firm based on Bandung, unveiled the Enrol Pilot in a December 2020 video that appeared to show the aircraft conducting flight tests. According to local media reports, work on the Enrol Pilot has been underway since 2017 and is supported by the Indonesian Ministry of Research and Technology. It can reportedly be equipped with 0.8kg (1.7lb) of explosive payload.

Specifications

Class	Loitering munition
Status	B
Unveiled	2020
Operators	
Length	0.8m (3ft)
Width (wingspan)	1.2m (4ft)
Height	
Maximum take-off weight	3kg (7lb)
Speed (max)	250km/h (135kts)
Range	30km (19 miles)
Endurance	20 hours
Flight ceiling	
Payload	0.8kg (2lb)
Armament	

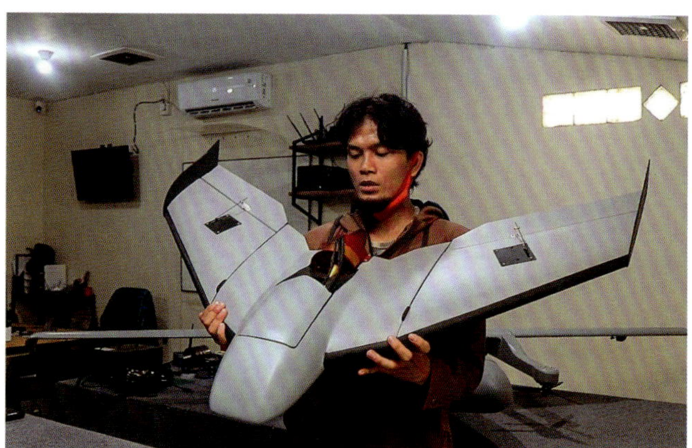

Enrol Sistem Pilot. (Kemenristek BRIN)

'Enrol Pilot – Inilah Drone Khusus Kamikaze Pertama Rancangan Indonesia', *Indomiliter*, 3 December 2020,
https://www.indomiliter.com/enrol-pilot-inilah-drone-khusus-kamikaze-pertama-rancangan-indonesia/

PTDI Elang Hitam

The Elang Hitam is a demonstrator Class III fixed-wing UCAV developed by a consortium led by PT Dirgantara Indonesia (PTDI). The Indonesian Ministry of Defense, Agency for Assessment and Application of Technology (BPPT), Indonesian Air Force, and the Institute of Aeronautics and Space are also involved in the project. The development of the Elang Hitam reportedly began in 2015. The PDTI-led consortium unveiled a full-scale mock-up of the Elang Hitam in a ceremony in December 2019. The consortium has repeatedly described the Elang Hitam as a strike-capable UAV. Local media reports have suggested that it could be equipped with PTDI's 5kg (11lb) Folding Fin Aerial Rocket (FFAR). It has low-mounted wings and a V-tail that is characteristic of many medium-altitude long-endurance UAVs. In January 2021, BPPT announced that flight trails would begin in August.

Specifications

Class	III
Status	C/C
Unveiled	2019
Operators	
Length	8.65m (28.4ft)
Width (wingspan)	16.0m (52ft)
Height	2.6m (9ft)
Maximum take-off weight	1,300kg (2,860lb)
Speed (cruise)	50 to 180km/h (27 to 97.2kts)
(maximum)	235km/h (126.9kts)
Range	
Endurance	24 hours
Flight ceiling	7,200m (24,000ft)
Payload	
Armament	(FFAR)

PDTI displayed a full-scale mockup of the Elang Hitam at a ceremony in 2019. (PDTI)

Maskur, F., 'PUNA MALE Elang Hitam Ditargetkan Terbang Perdana Agustus 2021', *Bisnis.com*, 28 January 2021,
https://ekonomi.bisnis.com/read/20210128/257/1349449/puna-male-elang-hitam-ditargetkan-terbang-perdana-agustus-2021

Wong, K., 'Indonesia reveals armed reconnaissance UAV development programme', *Janes*, 6 January 2020,
https://www.janes.com/defence-news/news-detail/indonesia-reveals-armed-reconnaissance-uav-development-programme

IRAN

HESA Ababil-2

The Ababil-2 is a family of loitering munitions developed by the Iran Aircraft Manufacturing Industrial Company (HESA). Although some variations exist among different users, the Ababil-2's standard layout features a cylindrical fuselage and delta wing design with top-mounted canards and a pusher propeller. It is believed to have a payload capacity of around 30 kg (66lb) and employs a rocket-assisted take-off system for launch and parachute for recovery. The Ababil-2 is an upgraded version of the Ababil-1, a retired loitering munition produced in the mid-1980s. It was unveiled at the 1999 IDEX defence exhibition. Initially produced as an aerial target for air defence training, the Ababil-2 has been modified to provide surveillance and attack capabilities. The attack version of the Ababil-2 is often referred to as the Ababil-T. The Ababil-2 is believed to be widely used by Iran's military and paramilitary forces. However, it is better known as one of Iran's most popular UAV exports; variants of the Ababil-2 are believed to be in use with at least three Iranian-backed militant groups in Lebanon, Palestine and Yemen. In 2004, Hezbollah dispatched an Ababil-2 – known locally as the Mirsad-1 – from Lebanon to Israel. More recently, the Ababil-2 has also been deployed by the Houthi group in Yemen as the Qasef-1 and by the Al Qassem Brigades in Gaza as the Shebab. A 2018 report by a U.N. panel of experts found that variants of the Ababil-2s used by the Houthi group in Yemen were virtually identical to those made by the Iran Aircraft Manufacturing Industrial Company.

Specifications

Class	Loitering munition
Status	A
Unveiled	1999
Operators	Iran, non-state groups
Length	2.5m (8.2ft)
Width (wingspan)	3.0m (9.8ft)
Height	
Maximum take-off weight	80kg (176lb)
Speed (max)	
Range	100km (62 miles)
Endurance	2 hours
Flight ceiling	
Payload	30kg (66lb)
Armament	

An HESA Abadil-2 on display. (Keyvan Tavakkoli)

Franchi, P. La, 'Iranian-made Ababil-T Hezbollah UAV shot down by Israeli fighter in Lebanon crisis', *FlightGlobal*, 15 August 2006,
https://www.flightglobal.com/iranian-made-ababil-t-hezbollah-uav-shot-down-by-israeli-fighter-in-lebanon-crisis/68992.article

United Nations, Security Council, Letter dated 26 January 2018 from the Panel of Experts on Yemen mandated by Security Council resolution 2342 (2017) addressed to the President of the Security Council, (S/2018/68) 26 January 2018, available from
https://reliefweb.int/sites/reliefweb.int/files/resources/N1800513.pdf

Wright, G., 'Ababil UAV', The Arkenstone (blog), 5 February 2011,
https://thearkenstone.blogspot.com/2011/02/ababil-uav.html

HESA Ababil-3

The Ababil-3 is a Class II fixed-wing UAV developed by Iran Aircraft Manufacturing Industrial Company (HESA). It features high-mounted wings, fixed tricycle landing gear, twin-booms with a mid-mounted horizontal stabilizer, and a push-propeller configuration. The Ababil-3 is part of HESA's family of Ababil UAVs, the first of which were developed in the 1980s. The Ababil-3 is believed to have entered military service with the Iranian Revolutionary Guard Corps in 2010, though it likely made its debut some years earlier. Since then, the Ababil-3 has largely been deployed in an unarmed capacity, providing part of the backbone of Iranian unmanned surveillance and reconnaissance capabilities. However, in an event in 2020, Iran's Ministry of Defence and Armed Forces Logistics pushed the first photographs of an armed Ababil-3. These upgraded, strike-capable Ababil-3 variants featured one hardpoint under each wing fitted with what appeared to be Ghaem-1 precision-guided bombs. A video published by Iranian state television in April 2020 showed the Ababil-3 taking-off with Ghaem-1 munitions. In January 2021, Iranian media outlets published video footage from an Iranian Army exercise of an Ababil-3 conducting a weapons trial with a new munition, the Almaz (الماس) air-to-ground guided missile. According to Iranian media reports, the Almaz missile has a range of 8km (4.9 miles) and is designed to target armoured vehicles. The Ababil-3 is in service with Iranian Revolutionary Guard Corps (IRGC). Variants of the Ababil-3 have also been deployed by Sudan and Venezuela.

Specifications

Class	II
Status	A/A
Unveiled	2010
Operators	Iran, Sudan, Venezuela
Length	4.5m (14.8ft)
Width (wingspan)	6.5m (21.3ft)
Height	
Maximum take-off weight	
Speed (maximum)	200km/h (108kts)
Range (operational)	250km (155 miles)
Endurance	8 hours
Flight ceiling	
Payload	
Armament	Almaz, Ghaem-1

An Ababil-3 on display, armed with what appear to be Almaz missiles. (Keyvan Tavakkoli)

BINNIE, J., 'Iran unveils armed Ababil-3 UAV', *Janes*, 20 April 2020,
https://www.janes.com/defence-news/news-detail/iran-unveils-armed-ababil-3-uav

LYAMIN, Y. (@imp_navigator), 'Thread on new UAVs drill of the Iranian Army that started on 5 Jan 2021', Twitter, 5 January 2021,
https://twitter.com/imp_navigator/status/1346398181924483074

WRIGHT, G., 'Ababil UAV', The Arkenstone (blog), 5 February 2011,
https://thearkenstone.blogspot.com/2011/02/ababil-uav.html

گشفتی نظامی ارتش مقر در خودرو/ موشک الماس از پهپاد ابابیل شلیک شد/ انجام عملیات پهپادی درب از بلند سواحل مکران,' *Khabar Online*, January 2021,
https://www.khabaronline.ir/news/1474355/

HESA Karrar

The Karrar is a Class II fixed-wing UCAV developed by the Iran Aircraft Manufacturing Industrial Company (HESA). The Karrrar features a long cylindrical fuselage and T-style wings. It bears a physical resemblance to target drones like the US MQM-107 or the South African HTD-1 Skua. Billed as Iran's first jet-powered UAV, Iran unveiled the Karrar in an August 2010 ceremony attended by then Iranian President Mahmoud Ahmadinejad. The Karrar was pictured fitted with a large bomb, possibly the 500lb (227kg) Mk 82. At the time, Iran claimed that the Karrar had a range of 1,000km (621 miles) and a maximum speed of 900km/h (486kts). Iran has said that the Karrar is compatible with various types of munitions. In a 2014 National Research Week exhibition of defence technology attended by Iranian President Hassan Rouhani, the Ministry of Defence displayed the Karrar with the 84kg (185lb) Shahab-e-Saqeb air-to-air missile. In November 2020, Iranian state-run media reported that a Karrar dropped a 500lb (227kg) bomb during a drill. In footage published by state-run media in 2021, a Karrar is seen launching an Azarakhsh air-to-air missile. The Karrar is believed to be in service with the Iranian military. In an April 2020 ceremony marking the introduction of several new UAVs, the Islamic Republic of Iran Army displayed the Karrar equipped with an unspecified munition.

Specifications

Class	II
Status	A/A
Unveiled	2010
Operators	Iran
Length	5m (18ft)
Width (wingspan)	3m (9.8ft)
Height	
Maximum take-off weight	
Speed (maximum)	900km/h (486kts)
Range	1,000km (621.4 miles)
Endurance	
Flight ceiling	
Payload	
Armament	Azarakhsh, Shahab-e-Saqeb, Mk 82 bomb

A Karrar during a test launch. (Keyvan Tavakkoli)

AL SALAMI, J., 'Iran's New Drone Is a Twin-Engine Bomber', War is Boring (blog), 3 February 2015,
https://medium.com/war-is-boring/irans-new-drone-is-a-twin-engine-bomber-6e5019f01828

NEWDICK, T., 'Iran Just Fired Its Sidewinder Missile Clone From a Drone', *The Drive*, 6 January 2021,
https://www.thedrive.com/the-war-zone/38580/iran-just-fired-its-sidewinder-missile-clone-from-a-drone

POMEROY, R., 'Iran unveils drone aircraft to counter 'aggressors'', *Reuters*, 22 August 2010,
https://www.reuters.com/article/us-iran-military-drone-idUSTRE67L0K120100822

RADIO FARDA, 'Iran Army Introduces New Drones, Upgrades Older UAV Models', *Radio Farda*, 21 April 2020,
https://en.radiofarda.com/a/iran-army-introduces-new-drones-upgrades-older-uav-models/30568207.html

RUBIN, M., 'A Short History of the Iranian Drone Programme', American Enterprise Institute, August 2020,
https://www.jstor.org/stable/pdf/resrep26682.pdf?refreqid=excelsior%3A1b6ccb30506c4a52e981951d5092683d

SYKES, P., 'Iran Uses Bomber Drones for First Time in Military Drones', *Bloomberg*, 3 November 2020,
https://www.bloomberg.com/news/articles/2020-11-03/iran-uses-bomber-drones-for-first-time-in-military-drills

HESA Shahed-123

The Shahed-123 is a Class II fixed-wing UAV developed by the Iran Aircraft Manufacturing Industrial Company (HESA). The Shahed-123 features a stubby fuselage with tricycle landing gear, high-mounted wings, and conventional V-tail. Work on the Shahed-123 began in the mid-2000s and HESA produced the first aircraft in 2007, according to researchers at Bellingcat. A Shahed-123 participated in the Great Prophet 7 military exercises in July 2012. Remains of Shahed-123s have been discovered in the disputed territory of Nagorno-Karabakh in 2011, Turkey in 2015, Iraq in 2016, and Afghanistan in 2016. Until recently, the Shahed-123 appeared to be unarmed and used solely for gathering intelligence. In a video released in March 2019 promoting the 'Towards Jerusalem' drone exercise, a Shahed-123 appears equipped with a precision-guided munition, perhaps the Sadid-345.

Specifications

Class	II
Status	A/A
Unveiled	2007
Operators	Iran
Length	
Width (wingspan)	
Height	
Maximum take-off weight	
Speed (maximum)	
Range	
Endurance	
Flight ceiling	
Payload	
Armament	Precision-guided munition

Shahed-123. (Keyvan Tavakkoli)

'Azerbaijani Drone Reportedly Downed Over Nagorno-Karabakh', *RFE/RL*, 14 September 2011,
https://www.rferl.org/a/karabakh_says_azerbaijani_drone_shot_down/24328599.html

Faraday, J. (@CT Studies), '#Iraq Amaq claims the downing of an Iranian made #UAV of the Iraqi Forces over Al-Qayyārah district south of #Ninawa', Twitter, 15 February 2016, https://twitter.com/ctstudies/status/699264689344274432

Gambrell, J., 'Devices found in missiles, Yemen drones link Iran to attacks', *Associated Press*, 19 February 2020,
https://abcnews.go.com/International/wireStory/devices-found-missiles-yemen-drones-link-iran-attacks-69064032

Rawnsley, A., 'Syria's 'New' Iranian Drone', *Bellingcat*, 28 January 2016,
https://www.bellingcat.com/news/mena/2016/01/28/syria-new-iranian-drone/

Shay, Dr. S., ''Towards Jerusalem': Iran's Large-Scale Drone Exercise', *IsraelDefense*, 17 March 2019,
https://www.israeldefense.co.il/en/node/37827

HESA Shahed-129

The Shahed-129 (شاهد 129) is a Class III fixed-wing UCAV developed by the Iran Aircraft Manufacturing Industrial Company (HESA). The Shahed-129 features a slender fuselage, high-mounted wings, and a conventional V-tail. In September 2013, Iran revealed that a prototype of the Shahed-129, its first medium-altitude long-endurance UAV, had conducted its first flight and that it could carry weapons. The following year, in September 2013, Iran displayed an attack version of the

Shahed-129 equipped with four Sadid-1 air-to-ground guided missiles on one hardpoint under each wing. Iran claimed at the time that the Shahed-129 could be equipped with up to eight munitions, though most photographs show it fitted with four or fewer. In a 2014 exhibition hosted for Iranian Supreme Leader Ali Khameni, the Iranian Revolutionary Guard Corps Air Force displayed a Shahed-129 with Sadid-1 missiles and Sadid-345 guided bombs. Beginning in 2014, Iran deployed armed Shahed-129s to Syria in support of the Assad regime, two of which were downed by the US Air Force in 2017. An armed Shahed-129 participated in the Great Prophet 9 wargame in February 2015. In 2019, the Iranian Navy announced that it had developed a maritime variant of the Shahed-129 known as the Simorgh. An Iranian Navy Shahed-129 conducted a weapons trial in September 2020 during the Zolfaghar-99 exercise. Today, the Shahed-129 is in active military service with the Iranian Navy and the Iranian Revolutionary Guard Corps Air Force.

Specifications

Class	III
Status	A/A
Unveiled	2012
Operators	Iran
Length	8.0m (26ft)
Width (wingspan)	16.0m (52ft)
Height	3.1m (10ft)
Maximum take-off weight	
Speed (cruise)	150km/h (81kts)
Range (LOS)	200km (124 miles)
Endurance	24 hours
Flight ceiling	7,300m (24,000ft)
Payload	400kg (880lb)
Armament	Sadid-1, Sadid-345

BBC STAFF, 'Iran unveils 'indigenous' drone with 2,000km range', *BBC*, 26 September 2012, https://www.bbc.com/news/world-middle-east-19725990

CENCIOTTI, D., 'Iranian Shahed 129 drone appears over Damascus', *The Aviationist*, 10 April 2014, https://theaviationist.com/2014/04/10/shahed-drone-over-syria/

'Iran Unveils New Naval Drone', *Tasnim News*, 7 December 2019, https://www.tasnimnews.com/en/news/2019/12/07/2154755/iran-unveils-new-naval-drone/

NERSISYAN, L., 'Drills reflect Iranian emphasis on unmanned and anti-ship capabilities', *Shephard News*, 29 September 2019, https://www.shephardmedia.com/news/naval-warfare/premium-drills-reflect-iranian-emphasis-unmanned-a/

RAWNSLEY, A., 'Drone War Heats Up in the Skies Over Syria', *Daily Beast*, 9 June 2017, https://www.thedailybeast.com/drone-war-heats-up-in-the-skies-over-syria?source=twitter&via=desktop/

The Shahed-129 with multiple Sadid-345 missiles on display. (Keyvan Tavakkoli)

IRIADF Kian

The Kian is a fixed-wing loitering munition. Also known as the Arash, the Kian was reportedly developed by Iran's Air Defence Forces. It features a cylindrical fuselage with mid-mounted wings and single vertical stabilizer. It employs a single engine in the pusher configuration and a rocket-assisted take-off system for launch. Iran has produced at least two variants of the Kian drone. The original Kian, now known as the Kian 1, is a jet-powered target drone first revealed in 2014 and used for training air defence forces. A Kian 1 target drone was on display in a May 2016 exhibition at Khatam al-Anbia Air Defence Base. In a September 2019 ceremony in Tehran, Brigadier General Alireza Sabahifard unveiled a larger, attack version of the Kian – the Kian 2 – claiming that it was designed to target enemy air defences and targets 'far from the country's borders'. The Arash appears to be a variant of the Kian 2, though the precise differences between the two are unclear; it made its first appearance in 2020. In a UAV exercise in January 2021, the Iranian military displayed 16 Kian/Arash loitering munitions. Around a dozen Kian/Arash loitering munitions were on display during an August 2021 visit by Iranian Army Air Force Brigadier General Aziz Nasirzadeh to Bandar Abbas Air Base.

Specifications

	Kian-2
Class	Loitering munition
Status	A
Unveiled	2014
Operators	Iran
Length	4.0m (13ft)
Width (wingspan)	3.5m (11.5ft)
Height	
Maximum take-off weight	
Speed (max)	480km/h (259kts)
Range	1,000km (620 miles)
Endurance	
Flight ceiling	5,000m (16,400ft)
Payload	
Armament	

IRIADF Kian. (Iranian media)

AL JAZEERA STAFF, 'Iran unveils high-precision reconnaissance and attack drone', *Al Jazeera*, 1 September 2019,
https://www.aljazeera.com/news/2019/9/1/iran-unveils-high-precision-reconnaissance-and-attack-drone

IRIB STAFF, 'فرهامده نیروی هوایی ارتش در بندرعباس؛ نیروی هوایی آمادهٔ دفاع تمام عیار', *IRIB News Agency*, 10 August 2021

MASHREGH NEWS STAFF, 'آزمایش "پهپاد جت کیان" برای اولین بار', *Mashregh News*, December 2014,
https://mshrgh.ir/374679

RADIO FARDA STAFF, 'ایران پس از جنگ سال بار دیگر از «پهپاد کیان» رونمایی کرد', *Radio Farda*, 21 September 2019,
https://www.radiofarda.com/a/Iran-unveils-KIAN-drone/30140582.html

ZWIJNENBURG, W., 'Persian Posturing: Iran's Drone Fleet Seen From Space', Bellingcat, 19 April 2021,
https://www.bellingcat.com/news/mena/2021/04/19/persian-posturing-irans-drone-fleet-seen-from-space/

Qods Mohajer-6

The Mohajer-6 is a Class II fixed-wing UCAV developed by Qods Aviation Industry Company. It features high-mounted wings, twin-booms with a high-mounted vertical stabilizer, tricycle landing gear, and a push-propeller configuration. The Mohajer-6 is part of the Mohajer family of midsized UAVs, the first of which, like the Ababil family, were developed in the mid-1980s. Iran's Ministry of Defence unveiled the Mohajer-6 in a ceremony in April 2017 that was attended by Iranian President Hassan Rouhani. Later that year, Qods Aviation displayed a subscale model of an armed Mohajer-6 at Russia's MAKS 2017 defence exhibition. Unlike other Iranian UAVs and earlier versions of the Mohajer that transitioned from an ISR to a combat role, Iran promoted the Mohajer-6 explicitly as a strike-capable UAV. It contains one hardpoint under each wing and is commonly associated with the Ghaem-series precision-guided bombs. In a February 2018 ceremony, the Mohajer-6 was exhibited fitted with two Ghaem-1s. In a September 2020 ceremony, the Iranian Revolutionary Guard Corps Navy displayed 15 Mohajer-6s, some of which were equipped with four Ghaem-1 bombs – two fitted under each wing. While the Iranian military remains the sole known operator of the Mohajer-6, it could be exported to Iranian allies in the future. In a ceremony in November 2020, the Venezuelan President Nicolas Maduro displayed a model of a UAV that resembled the Mohajer-6.

Specifications

Class	II
Status	A/A
Unveiled	2017
Operators	Ethiopia (reported), Iran
Length	5.67m (18ft)
Width (wingspan)	10.0m (32ft)
Height	
Maximum take-off weight	600kg (1,320lb)
Speed (maximum)	
Range (operational)	200km (124 miles)
Endurance	12 hours
Flight ceiling	
Payload	40kg (88lb)
Armament	Ghaem-1, Sadid-345

Mohajer-6 with Ghaem-1 missiles. (Keyvan Tavakkoli)

DOMBE, A. R., 'Iran Unveils New Tactical Drone', *Israel Defense*, 18 April 2017,
https://www.israeldefense.co.il/en/node/29263

DOMBE, A. R., 'Iran Begins Mass Production of Mohajer-6 Combat UAV', *Israel Defense*, 6 February 2018,
https://www.israeldefense.co.il/en/node/32964

KAMOZOV, V., 'Iranian Weapons and UAVs on Display at MAKS', *Aviation International Online*, 27 July 2017,
https://www.ainonline.com/aviation-news/defence/2017-07-27/iranian-weapons-and-uavs-display-maks

RUBIN, M., 'A Short History of the Iranian Drone Programme', American Enterprise Institute, August 2020,
https://www.jstor.org/stable/pdf/resrep26682.pdf?refreqid=excelsior%3A1b6ccb30506c4a52e981951d5092683d

LYAMIN, Y. (@imp_navigator), 'Among them IRGC Navy received 15 UCAVs Mohajer-6', Twitter, 23 September 2020,
https://twitter.com/imp_navigator/status/1308704147072483328/photo/1

SANCHEZ, W. A., 'Venezuelan technology plan may lean on Iran', *Shephard News*, 27 November 2020,
https://www.shephardmedia.com/news/uv-online/premium-venezuelan-technology-plan-may-lean-iran/

IAI Harpy, Harop and Mini Harpy

The Harpy is a large fixed-wing loitering munition developed by Israel Aerospace Industries. The Harpy features a delta wing design, a top-mounted canard front-plane, and a rocket-assisted booster. It is launched from a truck-mounted launcher, which can accommodate at least nine Harpy aircraft. Equipped with anti-radiation sensors to automatically detect and destroy hostile radar sites, the Harpy is designed primarily for the suppression of enemy air defence (SEAD) mission. IAI publicly revealed its existing in 1988, when it urged the US Air Force to evaluate the system. It is one of the earliest examples of a loitering munition. Between 1994 and 2003, Israel exported an unknown number of Harpy loitering munitions to China, which has produced its own variant of the system known as the ASN-301. It has also been exported to South Korea and Turkey and is believed to be in service with the Israeli military. In 2008, IAI revealed that it was developing a slightly larger successor to the Harpy known as the Harop, which was officially unveiled at the 2009 Paris Air Show. Unlike the Harpy, which operates autonomously within a pre-defined mission area, the Harop features a two-way datalink, enabling an operator to remotely direct the loitering munition to select targets and attack. In addition to Israel, the Harop is believed to be in service with Azerbaijan, India, Singapore and Turkey. It has been deployed by Azerbaijan repeatedly in conflicts with neighbouring Armenia over the disputed territory of Nagorno-Karabakh, most notably in 2020. In 2019, IAI unveiled the Mini Harpy, a lightweight variant that combines elements of the Harpy and Harop.

Specifications

	Harpy
Class	Loitering munition
Status	A
Unveiled	1988
Operators	China, India, Israel, South Korea
Length	2.1m (6.9ft)
Width (wingspan)	2.4m (7.9ft)
Height	
Maximum take-off weight	114kg (317lb)
Speed (maximum)	416km/h (224.6kts)
Range (operational)	150km (93 miles)
Endurance	9 hours
Flight ceiling	4,572m (15,000ft)
Payload (warhead)	32kg (70lb)
Armament	

Specifications

	Harop
Class	Loitering munition
Status	A
Unveiled	2009
Operators	Azerbaijan, Israel
Length	2.5m (8.2ft)
Width (wingspan)	3.0m (9.8ft)
Height	
Maximum take-off weight	
Speed (maximum)	416km/h (224.6kts)
Range (communication)	150km (93 miles)
Endurance	9 hours
Flight ceiling	4,572m (15,000ft)
Payload (warhead)	23kg (51lb)
Armament	

BEAL, S., 'Lessons from Nagorno-Karabakh hint at future procurement trends', *Shephard News*, 7 December 2020,
https://www.shephardmedia.com/news/landwarfareintl/premium-lessons-nagorno-karabakh-hint-future-procu/

DONALD, D., 'HAROP loiters with intent', *Aviation International Online*, 17 June 2009,
https://www.ainonline.com/aviation-news/defence/2009-06-17/harop-loiters-intent

FLIGHTGLOBAL STAFF, 'South Korea deploys anti-radar UAV', *FlightGlobal*, 17 January 2000,
https://www.flightglobal.com/south-korea-deploys-anti-radar-uav/30191.article

FLIGHTGLOBAL STAFF, 'USA and Israel in crisis over China Harpy deal', *FlightGlobal*, 3 January 2005,
https://www.flightglobal.com/usa-and-israel-in-crisis-over-china-harpy-deal-/58275.article

'IAI Urging US Air Force to Test Radar-tracking Drone', *Jewish Telegraphic Agency*, 8 April 1988,
https://www.jta.org/1988/04/08/archive/iai-urging-u-s-air-force-to-test-radar-tracking-drone

LAPPIN, Y., 'IAI announced new Mini Harpy loitering munition', *Janes*, 19 February 2019,
https://www.janes.com/defence-news/news-detail/iai-announces-new-mini-harpy-loitering-munition

Specifications

	Mini Harpy
Class	Loitering munition
Status	B
Unveiled	2019
Operators	
Length	
Width (wingspan)	
Height	
Maximum take-off weight	45kg (100lb)
Speed (max)	370km/h (200kts)
Range	100km (62 miles)
Endurance	2 hours
Flight ceiling	1,520m (5,000ft)
Payload	8kg (18lb)
Armament	

Harop. (IAI)

IAI Heron

The Heron is a family of Class III fixed-wing UAVs developed by Israel Aerospace Industries. The basic Heron design features a rectangular fuselage with bulging nose, high-mounted wings, and twin booms. The Heron employs a single engine in a pusher configuration and requires a runway for take-off and landing. The original Heron conducted its maiden flight in October 1994. Variants of the Heron include the Heron TP and Heron Mk II, which were unveiled in 2007 and 2020, respectively, as well as the Tactical Heron, smaller version of the Heron unveiled in 2019. Derivatives of the Heron include the Harfang, which was designed in collaboration with EADS to meet French military specifications and unveiled in 2003. The vast majority of Herons currently in operation are believed to be unarmed. As with other Israeli uninhabited combat aircraft, IAI has not publicly confirmed that the Heron can be armed. However, various media reports and non-governmental organizations have indicated that the Israeli military has deployed armed Herons equipped with Rafael Spike anti-tank guided missiles in conflicts in Lebanon and Palestine since 2006. Although Israel is believed to be the only military user of armed Herons, several other militaries – India and Germany – have considered weaponizing the aircraft. Media reports in 2020 and 2021 suggest that the Indian military may be close to a deal to upgrade its fleet of Herons with Spike missiles, though it is unclear as of this writing that it has done so.

Specifications

	Heron TP
Class	III
Status	A/A
Unveiled	1994 (Heron 1)
Operators	Israel
Length	14m (46ft)
Width (wingspan)	26m (85ft)
Height	
Maximum take off weight	5,670kg (12,474lb)
Speed (max)	407km/h (219.8kts)
Range	1,000km (620 miles)
Endurance	30 hours
Flight ceiling	13,716m (45,000ft)
Payload	2,700kg (5,940lb)
Armament	Spike

IAI Heron. (Boevaya mashina)

IAI Rotem

The Rotem is a rotary-wing loitering munition developed by Israel Aerospace Industries. It features a rectangular fuselage with four electric motors. The Rotem is hand-launched and available in a backpackable system that comes with two air vehicles and associated equipment. The Rotem's 1.2kg (3lb) warhead can be swapped for reconnaissance and surveillance payloads. IAI unveiled the Rotem at the 2016 Singapore Air Show. In 2021, IAI announced that an undisclosed international customer had awarded it a contract for an unspecified number of Rotem systems.

Specifications

Class	Loitering munition
Status	A
Unveiled	2016
Operators	
Length	
Width (wingspan)	
Height	
Maximum take-off weight	6kg (13lb)
Speed (max)	90km/h (48.6kts)
Range	10km (6 miles)
Endurance	0.5 hours
Flight ceiling	
Payload	1.2kg (3lb)
Armament	

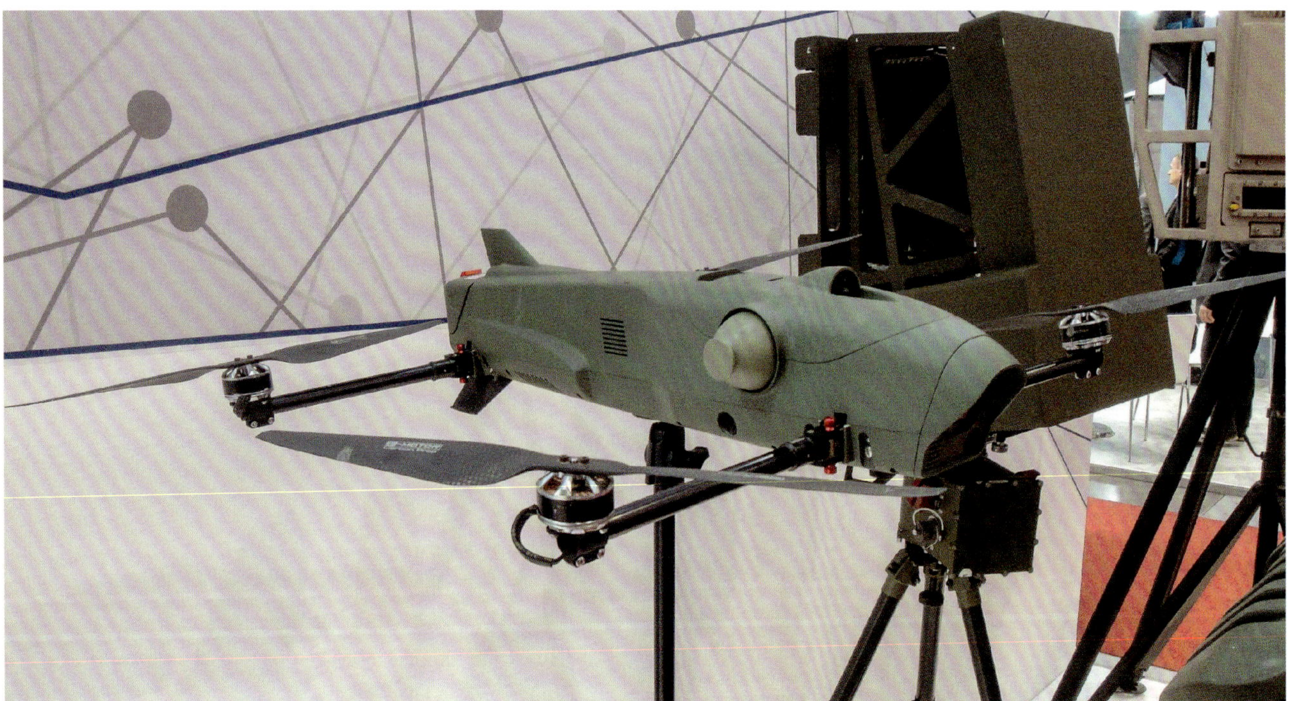

IAI Rotem. (Reise Reise)

Rafael Firefly

The Spike Firefly is a small rotary-wing loitering munition developed by Rafael. The Firefly features an 80mm (3.15in) fuselage and coaxial rotors that can fold into the body for storage. It is equipped with a 350g (0.8lb) omnidirectional fragmentation warhead. According to the manufacturer, the Firefly features include a dual seeker, homing algorithms, and computer vision, as well as a wave-off capability. A Firefly system for dismounted infantry is comprised of three air vehicles and a control unit, which together weigh a total of 15kg (33lb). Rafael unveiled the Firefly, which is part of the Spike family of precision munitions, in June 2018. In January 2019, Rafael conducted a series of tests in which it integrated the Firefly with various armoured vehicles. In May 2020, Rafael announced that it had been awarded a contract by the Israeli Ministry of Defense for an unspecified number of Firefly systems.

Specifications

	Firefly
Class	Loitering munition
Status	A
Unveiled	2018
Operators	Israel
Length	0.8m (2.6ft)
Width (wingspan)	
Height	
Maximum take-off weight	3.0kg (7.0lb)
Speed (maximum)	70km/h (37.8kts)
Range	1.0km (0.6 miles)
Endurance	0.25 hours
Flight ceiling	
Payload (warhead)	0.35kg (1.0lb)
Armament	

Firefly. (Rafael)

Ahronheim, A., 'FireFly: New variant in Rafael's Spike family of precision weapons', *Jerusalem Post*, 5 June 2018,
https://www.jpost.com/israel-news/firefly-new-variant-in-rafaels-spike-family-of-precision-weapons-559228

Cazalet, M., 'IAV 2020: Rafael integrates Firefly loitering munition with vehicles', *Janes*, 24 January 2020,
https://www.janes.com/defence-news/news-detail/iav-2020-rafael-integrates-firefly-loitering-munition-with-vehicle/

Hoyle, C., 'Israel orders Rafael's Spike Firefly loitering munition', *FlightGlobal*, 5 May 2020,
https://www.flightglobal.com/defence/israel-orders-rafaels-spike-firefly-loitering-munition/138229.article

Rafael, https://www.rafael.co.il/wp-content/uploads/2019/03/FIREFLY.pdf

Rafael, 'Spike 5th Gen Precision Guided Tactical Missiles', Rafael Advanced Defence Systems, accessed on 16 March 2021,
https://www.rafael.co.il/worlds/land/spike-5th-gen-precision-guided-tactical-missiles/

UVision Hero

The Hero is a family of fixed-wing loitering munitions developed by UVision. The Hero family encompasses nine different aircraft, which UVision has allocated to three groups – Tactical, Operational, and Strategic – according to weight class. The Tactical group includes the Hero-20, Hero-30, Hero-70, and Hero-120 aircraft. Each of these aircraft feature four fixed-planar cruciform wings and range from 3 to 12.5kg (6.6 to 28lb) in weight. Powered by electric engines and designed to be canister-launched, the Hero's Tactical loitering munitions are intended for short-range anti-personnel or light-armour missions. The Operational group includes the Hero-250, Hero-400, and Hero-400EC. The Hero-250 and -400 both feature a top-mounted wing and cruciform tail and are gasoline-powered, while the -400EC features a cruciform wing and tail and is powered by an electric engine. The Operational group systems have an endurance in excess of two hours and can be rail- or canister-launched. Finally, the Strategic group of Hero systems, the largest of the family, includes the Hero-900 and Hero-1250. Both feature top-mounted wings and a cruciform tail.

UVision unveiled the Hero-30 and Hero-400, the first two members of the family of loitering munitions, in 2013. Another four Hero systems – the Hero-70, -120, -250, and -900 – were unveiled at the 2015 Paris Air Show. The Hero-400EC was unveiled in 2017, the Hero-1250 in 2018, and the Hero-20 in 2019. UVision has moved aggressively to partner with foreign firms to market the Hero family to an international audience. In 2016, it partnered with Firstec to market the Hero-30 to the South Korean military and with Raytheon to market the Hero-30 to the US Army. In 2020, UVision partnered with Aditya Precitech to produce variants of the Hero family in India. In 2021, Northrop Grumman proposed a loitering munition based on the Hero-400 for the US Army's Air Launched Effects programme. UVision has also partnered with Estonia's MILREM robotics to integrate UVision's multi-canister launcher with the Type-X robotic combat vehicle, enabling it to carry the Hero-120 or Hero-400EC. While the Hero-30 is known to be in active military service with the Israel Defense Forces, the operational status of the other members of the Hero family is not clear. In 2019, UVision conducted a demonstration of the Hero-30 and Hero-400EC to an unnamed country in Asia and, in 2020, it completed a trial with an unspecified NATO-member navy. In 2021, the U.S. Marine Corps announced that it would acquire an undisclosed number of Hero-120s that will be integrated into armored ground vehicles and an autonomous boat. Also in 2021, Northrop Grumman announced that it would offer a variant of the Hero-400EC – the Hero ALE – for the U.S. Army's Air Launched Effects program, which seeks to develop drones and loitering munitions that can be launched from crewed rotary- and fixed-wing aircraft.

Specifications

	Hero-20	Hero-30	Hero-70	Hero-120	Hero-250	Hero-400
Class	Loitering munition	Loitering munition	Loitering munition	Loitering munition	Loitering munition	Loitering munition
Status	B	A	B	A	B	B
Unveiled	2019	2013	2015	2015	2015	2013
Operators		Israel		USA		
Length	0.64m (2.1ft)	0.8m (2.6ft)	1.0m (3.3ft)	1.34m (4.4ft)	1.8m (5.9ft)	2.2m (7.2ft)
Width (wingspan)	0.56m (1.8ft)	0.7m (2.3ft)	0.8m (2.6ft)	1.5m (4.9ft)	2.4m (7.9ft)	3.0m (9.8ft)
Height						
MTOW	1.8kg (4lb)	3.0kg (7lb)	7.0kg (15lb)	12.5kg (28lb)	25kg (55lb)	40kg (88lb)
Speed (maximum)						
Range (LOS)	10km (6 miles)	40km (25 miles)	40km (25 miles)	40km (25 miles)	40km (25 miles)	150km (93 miles)
Endurance	0.4 hours	0.5 hours	0.75 hours	1 hour	1 hour	4 hours
Flight ceiling						
Payload (warhead)	0.2kg (0.4lb)	0.5kg (1lb)	1.2kg (3lb)	4.5kg (10lb)	5.0kg (11lb)	8.0kg (18lb)
Armament						

Specifications

	Hero-400EC	Hero-900	Hero-1250
Class	Loitering munition	Loitering munition	Loitering munition
Status	B	B	B
Unveiled	2017	2015	2018
Operators			
Length	2.4m (7.9ft)	2.5m (8.2ft)	
Width (wingspan)	2.1m (6.9ft)	3.6m (11.8ft)	
Height			
Maximum take-off weight	40kg (88lb)	97kg (213lb)	120kg (275lb)
Speed (maximum)			
Range (LOS)	150km (93 miles)	250km (155 miles)	200km (124 miles)
Endurance	2 hours	7 hours	
Flight ceiling			
Payload (warhead)	10kg (22lb)	20kg (44lb)	30kg (66lb)
Armament			

Egozi, A., 'UVision reveals Hero-series UAS', *FlightGlobal*, 9 April 2013,
https://www.flightglobal.com/uvision-reveals-hero-series-uas/109334.article

Mönch Publishing Group, 'UVision Demonstrates HERO-30 and HERO-400EC to Asian Customer', 5 March 2019,
https://www.monch.com/mpg/news/land/5127-uvision-hero-asia.html

Mönch Publishing Group, 'UVision HERO-30 Successfully Completes a NATO Navy Trial', 23 June 2020,
https://www.monch.com/mpg/news/unmanned/7090-uvision-hero-30-successfully-completes-a-nato-navy-trial.html

Pocock, C., 'UVision Demonstrates Another Loitering Attack Drone', *Aviation International Online*, 11 January 2018,
https://www.ainonline.com/aviation-news/defence/2018-01-11/uvision-demonstrates-another-loitering-attack-drone

Reim, G., 'Northrop pitches UVision Hero loitering munition variant for US Army Future Vertical Lift', *FlightGlobal*, 10 March 2021,
https://www.flightglobal.com/military-uavs/northrop-pitches-uvision-hero-loitering-munition-variant-for-us-army-future-vertical-lift/142838.article

Reim, G., 'US Marine Corps to buy Uvision Hero-120 loitering munition', *FlightGlobal*, 21 June 2021,
https://www.flightglobal.com/military-uavs/us-marine-corps-to-buy-uvision-hero-120-loitering-munition/144245.article

Shephard News Team, 'Hero-30 enters service in Israel', *Shephard News*, 14 July 2020,
https://www.shephardmedia.com/news/air-warfare/hero-30-enters-service-israel/

Shephard News Team, 'Milrem and UVision join forces to offer new capability', *Shephard News*, 6 October 2020,
https://www.shephardmedia.com/news/uv-online/milrem-and-uvision-join-forces-offer-new-capabilit/

Trevithick, J., 'Northrop Grumman Reveals Sky Viper Chain Gun And New Suicide Drone For Future Helicopters', *The Drive*, 10 March 2021,
https://www.thedrive.com/the-war-zone/39696/northrop-grumman-reveals-sky-viper-chain-gun-and-new-suicide-drone-for-future-helicopters

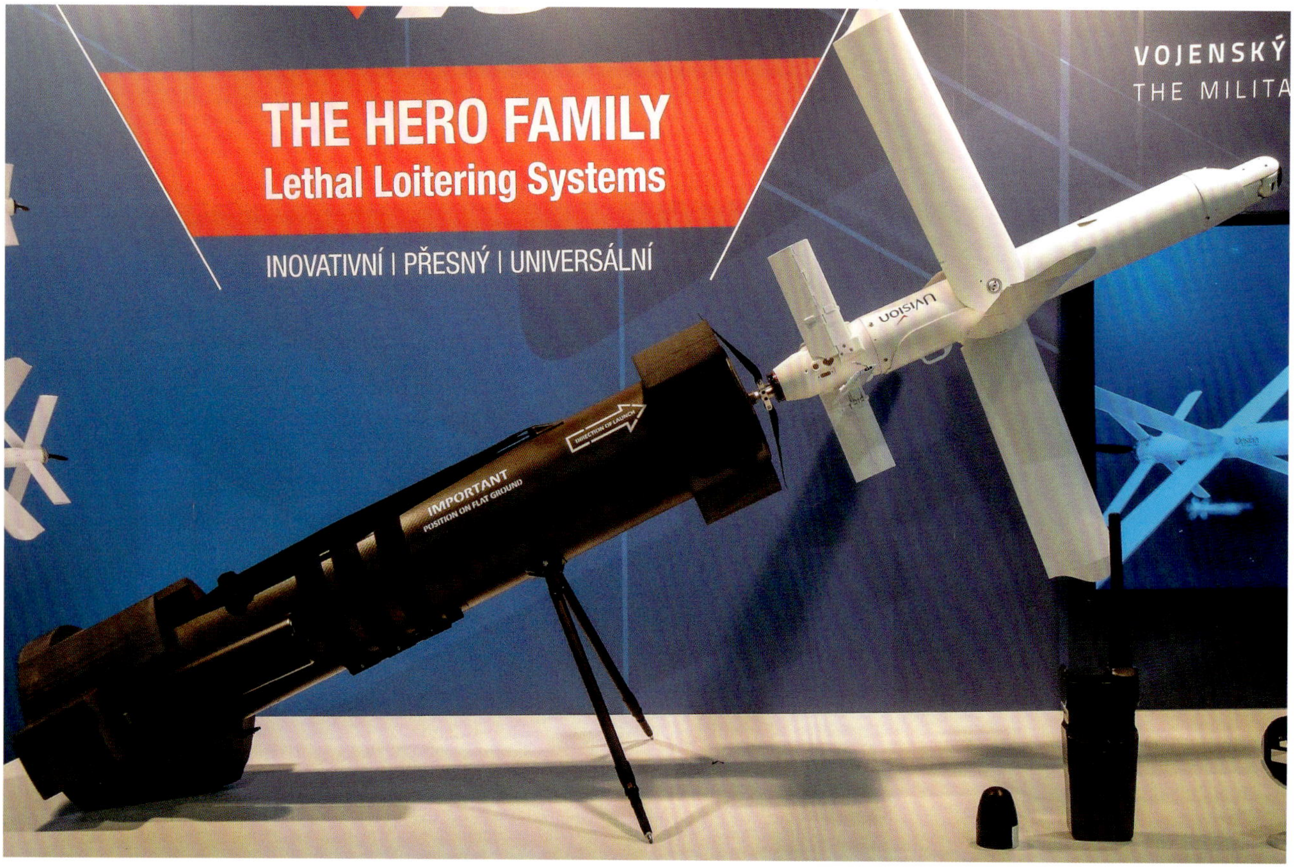

IAI Hero-30. (Reise Reise)

U-Visions Hero 400 and 250. (Georg Mader)

PAKISTAN

PAC MALE UAV

The Pakistan Aeronautical Complex (PAC) is in the process of developing an as yet unnamed Class III fixed-wing UCAV that is frequently referred to in marketing material as the medium-altitude long-endurance uninhabited aerial vehicle (MALE UAV). The MALE UAV project is led by the Aviation Research Indigenization and Development department at PAC. In 2017, the Pakistan Air Force revealed that efforts were underway at PAC to develop a Class III UAV as part of the Air Force's Project Azm modernization program. Various computer-generated images of the MALE UAV show it with two hardpoints under each wing, indicating that the system is designed to be capable of carrying out strikes. The project remains in development.

Specifications

Class	III
Status	C/C
Unveiled	2017
Operators	
Length	9m (29.5ft)
Width (wingspan)	15.8m (52ft)
Height	0.9m (3ft)
Maximum take-off weight	1,600kg (3,520lb)
Speed (max)	296km/h (159.8kts)
Range	
Endurance	
Flight ceiling	7,620m (25,000ft)
Payload	160kg (350lb)
Armament	

A subscale model of PAC's MALE UAV. (via Georg Mader)

QUWA STAFF, 'Pakistan Continues Investing in Drone Development', Quwa, 3 April 2021,
https://quwa.org/2021/04/03/pakistan-continues-investing-in-drone-development-2/

POLAND

MSP Inntech Warble

The Warble, or Warble Fly, is a fixed-wing loitering munition developed by MSP Inntech. The aircraft features a V-tail and foldable wings that extend after launch. According to MSP, it is compatible with a variety of payloads, including high explosive fragmentation and anti-tank fragmentation and thermobaric warheads, as well as inert and training warheads. MSP, a Polish firm that also produces target drones, unveiled the Warble in a September 2019 video. It was displayed at the IDEX 2021 defence exhibition alongside the AGEMA, a mission support unmanned ground vehicle developed by the UAE-based Milanion Group.

Specifications

Class	Loitering munition
Status	B
Unveiled	2019
Operators	
Length	1.35m (4.4ft)
Width (wingspan)	1.65m (5.4ft)
Height	
Maximum take-off weight	6.0kg (13lb)
Speed (maximum)	150km/h (81kts)
Range	10km (6 miles)
Endurance	0.5 hours
Flight ceiling	
Payload	
Armament	

MSP Warble (MSP)

MSP Design, 'Giez Warble loitering munition', YouTube, 10 September 2019,
https://www.youtube.com/watch?v=Dq-sKnptYPE

Shephard News Team, 'IDEX 2021: Milanion displays UGV with loitering munition and stabilized RWS', *Shephard News*, 22 February 2021,
https://www.shephardmedia.com/news/uv-online/idex-2021-milanion-displays-ugv-loitering-munition/

WB Group Warmate

The Warmate is a family of fixed-wing loitering munitions developed by WB Group that is currently comprised of the Warmate-1, Warmate-TL, Warmate-2, and Warmate-V. The Warmate-1, which is also known simply as the Warmate, can be equipped with a GO-1 warhead for unarmoured targets or GK-1 high-explosive anti-tank (HEAT) warhead. WB Group unveiled a mock-up of the Warmate at the 2014 MSPO defence exhibition. According to WB Group, the Warmate can conduct certain operations autonomously and serve as a man-portable alternative to anti-tank missiles. In November 2017, the Polish military awarded WB Group a contract for up to 1,000 Warmate loitering munitions, although the Warmate was not formally introduced until 2021 and it is not clear exactly how many of the loitering munitions have been delivered to date. WB Group has partnered with Ukraine's PJSC CheZaRa to produce warheads for the Warmate and with Germany's Rheinmetall to integrate the Warmate into the Rheinmetall Mission Master unmanned ground vehicle. The Warmate is also one element of the RSC Sokil, an integrated system of multiple armoured vehicles and drones produced by the Kyiv-based firm Ukrainian Armor. The Warmate may also have been used by an unknown actor in Libya in the spring of 2020.

In addition to the Warmate-1, WB Group has produced three other Warmate loitering munitions. The Warmate-TL is a tube-launched variant of the Warmate that features a slender fuselage with a foldable wing that expands upon launch. It is designed to be tube-launched either from a ground-based canister or from a vehicle. WB Group unveiled the Warmate-TL in 2019. The Warmate-2 is a fixed-wing loitering munition designed by WB Group and the UAE's Tawazun. WB Group unveiled the Warmate-2 at the MSPO 2018 expo. With a maximum take-off weight of 30kg (66lb) and a payload capacity of 5kg (11lb), the Warmate 2 is nearly six times heavier than the Warmate. It also has a slightly longer endurance and faster attack speed than its predecessor. Unlike the Warmate's man-portable tube-launch system, the Warmate 2 requires an elastomer launcher system, which can be mounted on a vehicle. According to WB Group, the Warmate 2 was designed to meet the specifications of the UAE Armed Forces, though it is unclear whether it is in active service with any military. The Warmate-V is a rotary-wing loitering munition. The Warmate-V features six coaxial motors on three arms. It can be equipped with a warhead weighing up to 1.6kg (3.5lb). WB Group unveiled the Warmate-V at Poland's MSPO 2019 defence exhibition. WB Group also produces a surveillance version of the Warmate called the Warmate-R.

Specifications

	Warmate–1	Warmate–TL	Warmate–2	Warmate–V
Class	Loitering munition	Loitering munition	Loitering munition	Loitering munition
Status	A	B	B	B
Unveiled	2014	2019	2018	2019
Operators	Poland			
Length	1.1m (3.6ft)	1.1m (3.6ft)	2.5m (8.2ft)	
Width (wingspan)	1.6m (5.2ft)	1.7m (5.6ft)	2.5m (8.2ft)	
Height				
MTOW	5.3kg (12lb)	4.5kg (10lb)	30kg (66lb)	7kg (15lb)
Speed (max)	150km/h (81kts)	120km/h (64.8kts)	160km/h (86.4kts)	72km/h (30.9kts)
Range	15km (9 miles)		20km (12 miles)	12km (7 miles)
Endurance	0.5 hours	0.5 hours	2 hours	0.5 hours
Flight ceiling	3,000m (9,800ft)		3,000m (9,800ft)	
Payload	1.4kg (3lb)		5.0kg (11lb)	1.6kg (4lb)
Armament				

Warmate-1. (WB Group)

Warmate-2. (WB Group)

Dombe, A. R., 'Poland Orders Thousand Suicide Drones', *Israel Defense*, 21 November 2017,
https://www.israeldefense.co.il/en/node/31823

Egozi, A. and Jarocki, M., 'Premium: Loitering munition finds in Libya highlight export control gaps', *Shephard News*, 24 April 2020,
https://www.shephardmedia.com/news/air-warfare/premium-loitering-munition-finds-libya-highlight-e/

Glowacki, B., 'Polish UAV producers target domestic sales', *FlightGlobal*, 10 September 2014,
https://www.flightglobal.com/military-uavs/polish-uav-producers-target-domestic-sales/114439.article

Interfax, 'Ukraine and Poland build up military-technical cooperation on program of Warmate mini drones', *Interfax-Ukraine*, 14 September 2019,
https://en.interfax.com.ua/news/general/448649.html

Jarocki, M., 'Poland formally commissions Warmate', *Shephard News*, 15 March 2021,
https://www.shephardmedia.com/news/air-warfare/premium-poland-formally-commissions-warmate/

Maundrill, B., 'MSPO 2018: Mighty Warmate 2 revealed', *Shephard News*, 5 September 2018,
https://www.shephardmedia.com/news/uv-online/mspo-2018-mighty-warmate-2-revealed/

Shephard News Team, 'Rheinmetall and WB Group introduce new version of Mission Master UGV', *Shephard News*, 4 September 2019,
https://www.shephardmedia.com/news/uv-online/rheinmetall-and-wb-group-introduce-new-version-mis/

WB Group, 'Warmate loitering munitions', accessed on 18 January 2021,
https://www.wbgroup.pl/en/produkt/warmate-loitering-munnitions/

WB Group, 'Warmate 2 loitering munitions', accessed on 18 January 2021,
https://www.wbgroup.pl/en/produkt/warmate-2-loitering-munitions-2/

WHITE, A., 'WB Group unveils the Warmate TL variant', *Janes*, 29 May 2019,
https://www.janes.com/defence-news/news-detail/wb-group-unveils-warmate-tl-variant

RUSSIA

Kalashnikov KUB-UAV

The KUB-UAV (Куб-БЛА) is a fixed-wing loitering munition produced by Kalashnikov Concern, a division of Rostec Corporation, and ZALA Aero Group. The KUB-UAV features a delta-wing design with a rounded fuselage and winglets. It employs a single electric engine in the pusher configuration and a catapult for launch. It was unveiled at the IDEX 2019 defence trade show. According to Kalashnikov, the KUB-UAV has a payload capacity of 3kg (6.6lb) and an endurance of 30 minutes. In December 2020, Rostec CEO Sergei Chemezov told reporters that the Russian military had used the KUB-UAV in Syria.

Specifications

Class	Loitering munition
Status	A
Unveiled	2019
Operators	Russia
Length	0.95m (3.1ft)
Width (wingspan)	1.2m (3.9ft)
Height	
Maximum take-off weight	19kg (42lb)
Speed (maximum)	130km/h (70.2kts)
Range	
Endurance	0.5 hours
Flight ceiling	
Payload	3.0kg (6.6lb)
Armament	

KUB-UAV. (Zala Aero Group)

Sʟʏ, L., 'The Kalashnikov assault rifle changed the world. Now there's a Kalashnikov kamikaze drone', *Washington Post*, 23 February 2019,
https://www.washingtonpost.com/world/2019/02/23/kalashnikov-assault-rifle-changed-world-now-theres-kalashnikov-kamikaze-drone/

ZALA Aᴇʀᴏ Gʀᴏᴜᴘ, 'KYB-UAV', accessed on 19 January 2021,
https://zala-aero.com/en/production/bvs/kyb-uav/

'Чемезов: БПЛА «Калашникова» успешно участвовали в боевых операциях в Сирии', *EurAsia Daily*, 7 December 2020,
https://eadaily.com/ru/news/2020/12/07/chemezov-bpla-kalashnikova-uspeshno-uchastvovali-v-boevyh-operaciyah-v-sirii

Kronstadt Grom

The Grom (Гром) is a concept Class III fixed-wing UCAV developed by Kronstadt Group, a subsidiary of Joint-Stock Financial Corporation Sistema. Kronstadt unveiled a mock-up of the Grom at the Army-2020 defence exhibition in Kubinka. In an interview with Russian media outlet Interfax, Vladimir Voronov, Kronstadt's research director, said that the Grom was designed to operate as a loyal wingman for manned aircraft. Voronov specifically suggested that the Grom could be used to suppress enemy air defences and radar stations. Powered by two AI-222-25 turbofan engines, the Grom could reach speeds of up to 1,000km/h (540kts). Kronstadt displayed the Grom at Army-2020 alongside the KAB-250-LG-E, KAB-500S-E, and X-38MLE munitions. According to Voronov, the Grom will have two external hardpoints, one under each wing, as well as an internal weapons bay. In March 2021, a Kronstadt spokesperson claimed that the Grom would be capable of launching Molniya drones, a loitering munition currently under development.

Specifications

Class	III
Status	C/C
Unveiled	2020
Operators	
Length	13.8m (45ft)
Width (wingspan)	10.0m (33ft)
Height	3.8m (12ft)
Maximum take-off weight	7,000kg (15,44lb)
Speed (maximum)	1,000km/h (540kts)
Range	700km (434 miles)
Endurance	
Flight ceiling	12,000m (40,000ft)
Payload	2,000kg (4,409lb)
Armament	(KAB-250-LE-E), (KAB-500S-E), (X-38MLE)

Kronstadt Grom. (via Twitter)

Butowski, P., 'Russia Reveals Loyal Wingman Concept', *Aviation Week*, 4 September 2020,
https://aviationweek.com/special-topics/air-dominance/russia-reveals-loyal-wingman-concept

Stefanovich, D. (@KomissarWhipla), 'Grom high-speed strike UAV by Kronstadt group', Twitter, 28 August 2020,
https://twitter.com/KomissarWhipla/status/1299272397682626562/photo/3

'Скоростной ударный стелс-беспилотник 'Гром' способен прорвать и уничтожить ПВО противника', *Interfax-AVN*, 25 August 2020,
https://www.militarynews.ru/story.asp?rid=1&nid=536847&lang=RU

Kronstadt Helios

The Helios (Гелиос) is a concept Class III fixed-wing UAV developed by Kronstadt Group, a subsidiary of Joint-Stock Financial Corporation Sistema. Kronstadt unveiled a mock-up of the Helios at the Army-2020 defence exhibition. The Helios features a push-propeller powerplant, mid-mounted wings, and a twin-boom and inverted V-tail design. The mock-up included an early-warning large surveillance radar equipped under the fuselage. As of this writing, the Helios has yet to conduct a flight test.

Specifications

Class	III
Status	C/C
Unveiled	2020
Operators	
Length	12.6m (41ft)
Width (wingspan)	30.0m (98ft)
Height	
Maximum take-off weight	4,000kg (8,800lb)
Speed (cruise)	350–450km/h (189–242kts)
Range (BLOS)	3,000km (1,860 miles)
Endurance	24–30 hours
Flight ceiling	9,000m (30,000ft)
Payload	800kg (1,760lb)
Armament	

Kronstadt Helios. (Russian Ministry of Industry and Trade)

Butowski, P., 'Russia Reveals Loyal Wingman Concept', *Aviation Week*, 4 September 2020,
https://aviationweek.com/special-topics/air-dominance/russia-reveals-loyal-wingman-concept

Lee, R., 'And the Helios UCAV. 116', Twitter, 25 August 2020,
https://twitter.com/RALee85/status/1298240798706860032/photo/2

Kronstadt Molniya

The Molniya (Молния) is a conceptual loitering munition under development by Kronstadt Group, a subsidiary of Joint-Stock Financial Corporation Sistema. At the Army-2020 defence exhibition, Kronstadt revealed that it was working on swarming drone system called the Piranha (Пиранья) modelled on the US Gremlins and LOCUST programmes. In early March 2021, Russian media outlets began to tease some of the physical characteristics and nominal specifications for the drone, such as that it would be jet-powered and could be launched from an inhabited or uninhabited aircraft such as the Su-57 fighter or Hunter UCAV. In mid-March 2021, a Kronstadt spokesperson announced that the system previously known as Piranha would be called the Molniya (Lightning). The spokesperson also claimed that the Grom (Thunder), a conceptual loyal-wingman UCAV under development by Kronstadt, would be capable of equipping with 10 Molniya drones, which could operate in a swarm to carry out attack or reconnaissance missions.

Specifications

Kronstadt Molniya. (Russian Ministry of Industry and Trade)

Class	Loitering munition
Status	C
Unveiled	2020
Operators	
Length	1.5m (4.9ft)
Width (wingspan)	1.2m (3.9ft)
Height	
Maximum take-off weight	
Speed (maximum)	700km/h (378kts)
Range	
Endurance	
Flight ceiling	
Payload (warhead)	7.0kg (15lb)
Armament	

Khodarenok, M., 'Пулеметы, дроны, истребители: что нового покажут на «Армии-2020»', *Gazeta.ru*, 23 August 2020,
https://www.gazeta.ru/army/2020/08/23/13207327.shtml

RIA staff, 'Источник: для ВКС создают работающие в стае реактивные беспилотники', *RIA.Ru*, 1 March 2021,
https://ria.ru/20210301/bespilotniki-1599368302.html

TASS staff, 'Russia's latest combat drone to control swarm of reconnaissance UAVs', *TASS*, 15 March 2021,
https://tass.com/defence/1265961

Kronstadt Orion

The Orion (Орион) is a Class III fixed-wing UCAV developed by Kronstadt Group, a subsidiary of Joint-Stock Financial Corporation Sistema. The Orion project began in 2011, when the Russian Ministry of Defence awarded Transas, the St Petersburg-based company that preceded Kronstadt, a contract to develop a medium-altitude long-endurance UAV. Kronstadt revealed the Orion in a video at the 2017 MAKS defence exhibition; photos of the aircraft were released in March 2018. With a 16-meter-long wingspan, the Orion is among the smaller MALE UAVs and resembles the AVIC Wing Loong I or GA-ASI Predator. It features mid-mounted wings, a slender fuselage, and V-tail. The MoD designation for the Orion is the Inokhodets (Иноходец). Kronstadt has conducted flight tests of the Orion at Protasovo Airport, where one Orion crashed during flight tests in November 2019. The Inokhodets (Иноходец) is a variant of the Orion produced to meet Russian Ministry of Defence requirements and not intended for export.

In September 2018, Kronstadt unveiled the Orion-E, an export-ready variant of the Orion that can carry 50kg (110lb) guided missiles. In August 2020, Kronstadt displayed the Orion at the Army-2020 defence exhibition alongside the KAB-20, KAB-50, and FAB-50 guided bombs, the Kh-50 guided missile, and the UPAB-50 glide bomb. In early December 2020, Russian Defence Minister Sergei Shoigu announced that the Aerospace Forces had taken delivery of the first Orion systems. Later that month, Alexei Krivoruchko, the deputy defence minister for armaments, said that the MoD would accelerate the acquisition of UCAVs in the following year. Krivoruchko also noted that Russia would begin producing small UAV-specific munitions in 2021. Also in December 2020, the MoD published an image of the Orion equipped with KAB-20 guided bombs. In February 2021, a Russian television program featured video footage of an armed Orion equipped conducting air operations in Syria. According to the broadcast, 17 of the 38 Orion sorties in Syria involved an airstrike. Though the footage of the Orion obscured the images of the munitions, the aircraft appeared to be equipped with two lightweight munitions under each wing and one unguided bomb – perhaps the FAB-50 – attached underneath the fuselage. In a July 2021 interview, Kronstadt CEO Sergei Bogatikov said that the Orion had conducted tests with the OFAB-100 fragmentation bomb. At the Army-2021 defence exhibition, Bogatikov told Russian media outlets that efforts to integrate guided missiles on the Orion were underway; tests with the Vikhr-M guided anti-tank missile were scheduled to occur by the end of the same year.

Specifications

Class	III
Status	A/A
Unveiled	2017
Operators	Russia
Length	8.0m (26ft)
Width (wingspan)	16m (52ft)
Height	
Maximum take-off weight	1,200kg (2,640lb)
Speed (maximum)	200km/h (108kts)
Range (operational)	300km (186 miles)
Endurance	24 hours
Flight ceiling	7,500m (24,600ft)
Payload	200kg (440lb)
Armament	KAB-20, OFAB-100, (Vikhr-M)

Kronstadt Orion. (Nickel Nitride)

Binnie, J., 'Russia announces Orion UAV sale to Middle East country', *IHS Jane's Defence Weekly*, 21 August 2018,
https://www.janes.com/article/82491/russia-announces-orion-uav-sale-to-middle-east-country

Lee, R. (@RALee85), https://twitter.com/RALee85/status/1045374311564029960

'Минобороны впервые показало беспилотник 'Орион' в ударном варианте и с вооружением', *Tass*, 28 December 2020,
https://tass.ru/armiya-i-opk/10365187

Newdick, T., , 'Russia Provides a Glimpse of its Orion Drone Executing Combat Trials in Syria', *The Drive*, 22 February 2021,
https://www.thedrive.com/the-war-zone/39381/russia-provides-a-glimpse-of-its-orion-drone-executing-combat-trials-in-syria

Novichkov, N., , 'Kronstadt weaponises Orion-E UAV, outlines HALE UAV development', *Jane's International Defence Review*, 26 September 2018,
https://www.janes.com/article/83350/kronshtadt-weaponises-orion-e-uav-outlines-hale-uav-development

'Российский беспилотник впервые применил управляемые ракеты', *RNA Novosti*, 28 December 2020,
https://ria.ru/20201228/bespilotnik-1591191802.html

Redstar.gr, 'MALE UAV 'Orion'',
https://www.redstar.gr/index.php?option=com_content&view=article&id=5273&catid=783&Itemid=530&lang=en

'Russian company release photos of new Orion UAV', *Defence Blog*, 26 March 2018,
https://defence-blog.com/news/russian-company-release-photos-new-orion-uav.html

TASS, 'Orion combat drone crashes near apartment house in Ryazan Region - investigators', 16 November 2019,
https://tass.com/emergencies/1089575

Kronstadt Sirius

The Sirius (Сириус) is a concept Class III fixed-wing UCAV developed by Kronstadt Group, a subsidiary of the Joint-Stock Financial Corporation Sistema. Kronstadt announced at the MAKS-2019 defence exhibition that it was developing a larger variant of the Orion known as Sirius. It unveiled a mock-up of the Sirius at the Army-2020 defence exhibition. Like its predecessor, the Sirius has mid-mounted wings, a narrow fuselage, and a V-tail. Unlike the Orion, the Sirius has two engines, one mounted on each wing. Kronstadt displayed the Sirius at Army-2020 alongside the FAB-500M-62, KB-100, OFAB-250, OFAB-500, and PNK-500U bombs. However, with a reported payload capacity of 450kg (992lb), the Sirius is expected to be equipped with munitions weighing up to 250kg (551lb). As of this writing, the Sirius has yet to conduct a flight test or weapons trial. In an interview with Russian media outlet Interfax, Vladimir Voronov, Kronstadt's research director, said that the Sirius would also be equipped with a satellite communications capability, extending its operational range. At the Army-2021 defence exhibition, the Russian Ministry of Defence announced that the Sirius would enter military service in 2023 as the Inokhodets-RU.

Specifications

Kronstadt Sirius. (Kronstadt)

Class	III
Status	C/C
Unveiled	2020
Operators	
Length	9.0m (29.5ft)
Width (wingspan)	23.0m (75ft)
Height	3.3m (11ft)
Maximum take-off weight	2,500kg (5,500lb)
Speed (cruise)	180km/h (97.2kts)
Range	
Endurance	20 hours
Flight ceiling	7,000m (23,000ft)
Payload	450kg (992lb)
Armament	

Butowski, P., 'Russia Reveals Loyal Wingman Concept', *Aviation Week*, 4 September 2020,
https://aviationweek.com/special-topics/air-dominance/russia-reveals-loyal-wingman-concept

Stefanovich, D. (@KomissarWhipla), 'Sirius recon-strike UAV by Kronstadt group', Twitter, 28 August 2020,
https://twitter.com/KomissarWhipla/status/1299272445296283648

'Скоростной ударный стелс-беспилотник 'Гром' способен прорвать и уничтожить ПВО противника', *Interfax-AVN*, 25 August 2020,
https://www.militarynews.ru/story.asp?rid=1&nid=536847&lang=RU

SAT AR-10 Argument

The AR-10 Argument (Аргумент) is a concept Class III fixed-wing UCAV designed by Modern Aviation Technologies (SAT). The AR-10 Argument based on the manned SR-10, a two-seat trainer aircraft also developed by SAT. Like the SR-10, the design for the SR-10 has forward-swept wings. The AR-10, however, features V-tail vertical stabilizers instead of the single vertical and two horizontal stabilizers on the ST-10. At the MAKS 2017 defence exhibition, SAT announced that the SR-10 would serve as the basis for a new UCAV called the AR-10 Argument that would be designed to conduct strikes and target enemy aircraft. It also displayed a computer-generated image of the planned AR-10 and preliminary specifications. In February 2021, posts to Russian social media revealed additional details about the AR-10, including that it will have a 4.4-meter (14.4ft)-long internal weapons bay that could accommodate the UKVU-50 ejection release unit and the X-38 air-to-ground missile. The AR-10 would have additional external hardpoints on the wings. The status of the project, however, is not clear as of this writing as it does not appear that a demonstrator aircraft has been produced.

Specifications

Class	III
Status	C/C
Unveiled	2017
Operators	
Length	9.7m (31.8ft)
Width (wingspan)	8.4m (27.6ft)
Height	3.4m (11.0ft)
Maximum take-off weight	3,850kg (8,470lb)
Speed (maximum)	910km/h (491.4kts)
Range	950km (589 miles)
Endurance	6 hours
Flight ceiling	11,500m (37,720ft)
Payload	680kg (1,500lb)
Armament	(X-38)

SAT AR-10 Argument.

BMPD staff, 'Проект ударного беспилотного летательного аппарата АР-10 «Аргумент»', BMPD (blog), 29 February 2021, https://bmpd.livejournal.com/4266580.html#cutid1

Karnozov, V., , 'Surprise Debut for Russian Light Jet at MAKS', *Aviation International Online*, 26 July 2017, https://www.ainonline.com/aviation-news/defence/2017-07-26/surprise-debut-russian-light-jet-maks

Sukhoi S-70 Okhotnik

The S-70 Okhotnik (Охотник), also known as the Hunter-B, is a demonstrator Class III UCAV developed by Sukhoi Company, a division of the United Aircraft Corporation. Russia's Ministry of Defence awarded Sukhoi a contract to begin developing the Okhotnik in 2011. The first images of the Okhotnik were published to social media in January 2019, revealing an aircraft with a low-observability, flying-wing design. The Okhotnik is reported in media sources to have a maximum take-off weight of around 20 metric tons, a length of 14m (45.9ft), and a wingspan of 19m (62.3ft). In August 2019, the Ministry of Defence announced that the Okhotnik had conducted its first flight that month. In September 2019, the MoD published a video showing the Okhotnik conducted a flight test alongside a manned Su-57. Each of the Okhotnik's flight tests to date have taken place at the 929th Chkalov State Flight Test Centre in Astrakhan. In a meeting with Russian President Vladimir Putin in early August 2020, Yury Slyusar, then the director of United Aircraft Corporation, said that the UAC would begin delivering the Okhotnik to the military beginning in 2024 in order to meet an accelerated timeline proposed by the Ministry of Defence. In the meantime, Sukhoi plans to produce two more Okhotnik prototypes in 2021 and 2022.

In his August 2020 meeting with Putin, UAC Director Yury Slyusar asserted that the Okhotnik would have the capacity to carry an exceptionally diverse collection of munitions and equipment compared to similar UCAVs. According to Russian media reports in late 2020 and early 2021, the Okhotnik has conducted a series of weapons trials involving both inert air-to-air missiles and a 500kg (1,102lb) free-fall unguided bomb. In these trials, the Okhotnik reportedly launched the weapons from one of its two internal weapons bays (внутрифюзеляжными). In a February 2021 interview with Russian media, Boris Obnosov, CEO of the Tactical Missiles Corporation, revealed that the Okhotnik had tested the Grom-E (Гром-Э), a guided missile based on the Soviet-era Kh-38ME air-to-surface missile. Follow-on weapons trials in 2021 are expected to involve air-to-surface guided missiles. Some defence analysts have suggested that the Okhotnik could serve in a loyal wingman (верным ведомым) capacity for manned aircraft, potentially the Su-57 fighter or Tu-95MS strategic bomber. Given that the Okhotnik remains in the early stages of flight and weapons trials, the ultimate configuration and mission of the aircraft could evolve in the years before it is expected to enter active service.

Specifications

Class	III
Status	B/B
Unveiled	2019
Operators	
Length	14m (49ft)
Width (wingspan)	19m (52ft)
Height	
Maximum take-off weight	20,000kg (44,400lb)
Speed (maximum)	1,000km/h (540kts)
Range	
Endurance	
Flight ceiling	18,000m (59,000ft)
Payload	
Armament	Grom-E

The S-70 during taxi trials during winter 2018-19. (Sukhoi)

UZGA Altair

The Altair (Альтаир), which also known as Altius (Альтиусом), is a demonstrator Class III fixed-wing UCAV. Simonov Design Bureau (SDB) in Kazan initiated work on the Altair in 2011 and produced two prototypes of the aircraft. The Altair conducted its first flight tests in 2016. The Altair was intended to be used for arctic missions and is notable for its twin-engine design. In October 2018, the Ministry of Defence pulled the project from SDB, noting a lack of progress and technical challenges in key areas such as sensors and microprocessors. In December 2018, Deputy Defence Minister Alexei Krivoruchko said in an interview with a Russian press outlet that work on the Altius would continue at the Ural Civil Aviation Plant (UZGA). According to a February 2020 report in the with Russian media outlet Izvestiya, the Tactical Missiles Corporation (Тактическое ракетное вооружение), a Russian defence firm, has developed a family of precision glide-bombs for the Altius and the Okhotnik UCAVs. This family includes the 9-A-7759, a 600kg (1,323lb) bomb with a range of 120km (75 miles), as well as two variants, the 9-A1-7759 and 9-A2-7759, each of which feature different warheads. In a June 2020 visit to Kazan by Deputy Defence Minister Alexei Krivoruchko, a Russian television newscast captured video of the Altair equipped with two large bombs, one on hardpoint under each wing, in what was the first visual confirmation of plans to arm the Altair. Although the type of the munition was not clear from the image, they did not appear to resemble those developed by the Tactical Missiles Corporation. In June 2021, Russian media sources reported that the Altair had conducted a series of weapons trials with various unspecified bombs and missiles.

Specifications

The Altair in flight. (Russian Ministry of Defence)

Class	III
Status	B/B
Unveiled	2016
Operators	
Length	11.6m (38ft)
Width (wingspan)	28.5m (93ft)
Height	
Maximum take-off weight	6,000kg (13,200lb)
Speed (max)	
Range	
Endurance	24 hours
Flight ceiling	12,000m (40,000ft)
Payload	1,500kg (3,300lb)
Armament	

GTRK Tatarstan, 'Казанский авиазавод посетил заместитель министра обороны страны Алексей Криворучко', YouTube, 20 June 2020, https://www.youtube.com/watch?v=MP1qeAXxEkw

Lavrov, A. and Ramm, A., '«Гром» победы: ударные дроны уничтожат ракетные комплексы противника', *Izvestiya*, 27 February 2020, https://iz.ru/975885/anton-lavrov-aleeksei-ramn/grom-pobedy-udarnye-drony-unichtozhat-raketnye-kompleksy-protivnika

RIA STAFF, "Ударный беспилотник 'Альтиус' впервые применил оружие, сообщил источник', *RIA Novosti*, 25 June 2021, https://ria.ru/20210625/altius-1738521644.html

'В Казани закрыта программа производства тяжелого беспилотника', *Federal Press*, 9 October 2018, http://fedpress.ru/news/16/economy/2139659

'Разработка беспилотника 'Альтиус' продолжится с новым подрядчиком', *TASS*, 19 December 2018, https://tass.ru/armiya-i-opk/5931392

UZGA Forpost-M/R

The Forpost (Форпост-M/R) is a Class II fixed-wing UCAV developed by the Ural Civil Aviation Plant (UZGA). It is a Russian derivative of the Israel Aerospace Industries Searcher Mk II, unarmed versions of which Russia has operated for reconnaissance and surveillance since the early 2010s. UZGA has developed two variants of the Forpost – the Forpost-M and Forpost-R – which are believed to be capable of conducting strikes. The Forpost-M and Forpost-R were announced at the Army-2015 arms fair. Of the two, however, only the Forpost-R is believed to be in active military service. The Forpost-R is outfitted with a new domestically produced engine – the APD-85 – and an upgraded electro-optical system, the GOES-540. In December 2019, the Russian Ministry of Defence announced that the Forpost-R had conducted its maiden flight. In early December 2020, Russian Defence Minister Sergei Shoigu announced that the Russian Aerospace Forces would begin taking delivery of strike-capable Forpost systems. At the Army defence exhibition in August 2021, UZGA displayed the Forpost-R equipped with an unspecified type of air-to-surface missile on hardpoints under each wing. The Russian MoD told media outlets at the time that two Forpost-R systems comprising of a total of six aircraft were already in service with the military. Also in August 2021, photos posted to social media appeared to show the wreckage of a downed Forpost-R in Syria's Aleppo province.

Specifications

Class	II
Status	A/A
Unveiled	2015
Operators	Russia
Length	5.9m (19ft)
Width (wingspan)	8.6m (28ft)
Height	1.4m (5ft)
Maximum take-off weight	500kg (1,100lb)
Speed (maximum)	204km/h (110.2kts)
Range	250km (550 miles)
Endurance	18 hours
Flight ceiling	5,900m (19,352ft)
Payload	100kg (220lb)
Armament	KUB-20

Aex Ru Staff, 'Разведывательный беспилотник "Форпост-Р" начинает госиспытания', *Aviation Explorer*, 23 December 2019, https://www.aex.ru/news/2019/12/23/206369/

Tass Staff, 'Новейшая ударная версия беспилотника 'Форпост' была показана на 'Армии-2021', *Tass*, 24 August 2021, https://tass.ru/armiya-i-opk/12203859

Steele, M. (@SteeleSyAA), 'Pictures of a damaged drone and crash site near the village of "Kafr Noren" in western #Aleppo province', Twitter, 4 August 2021, https://twitter.com/SteeleSyAA/status/1422776770206396419

A Forpost in use with the Baltic Fleet. (Russian Ministry of Defence)

ZALA Lancet

The Lancet is a family of fixed-wing loitering munitions produced by the Zala Aero Group, a subsidiary of Kalashnikov, and comprised of the Lancet-1 and Lancet-3. The Lancet features a cruciform design with four wings and four stabilizers. It employs a single electric motor in a pusher propeller configuration and a catapult launch system. Zala unveiled the Lancet-1 at the 2019 Army defence exhibition. In December 2020, a Russian media report claimed that Russian military forces in Syria had tested the capabilities of the Lancet and the KYB, another Russian loitering munition, in combat. Zala Aero unveiled the Lancet-3, a larger variant with greater payload capacity, at the 2021 Army defence exhibition. Speaking at the event, Zala Aero representative Alexander Zakharov suggested that Lancet loitering munitions could be used to target adversary uninhabited aircraft using a practice he referred to as 'air mining'. Zala Aero has also reportedly conducted a series of test launches of the Lancet from combat vessels.

Specifications

	Lancet-1
Class	Loitering munition
Status	A
Unveiled	2019
Operators	Russia
Length	
Width (wingspan)	
Height	
Maximum take-off weight	5.0kg (11lb)
Speed (max)	110km/h (59.4kts)
Range	40km (25 miles)
Endurance	0.5 hours
Flight ceiling	
Payload	1.0kg (2lb)
Armament	

Specifications

	Lancet-3
Class	Loitering munition
Status	B
Unveiled	2021
Operators	
Length	
Width (wingspan)	
Height	
Maximum take-off weight	12kg (26lb)
Speed (max)	110km/h (59.4kts)
Range	40km (25 miles)
Endurance	0.6 hours
Flight ceiling	
Payload	3.0kg (7lb)
Armament	

Zala Lancet. (Zala)

SAUDI ARABIA

KACST Saker-1

The Saker-1 is a series of prototype Class III fixed-wing UCAVs developed by the King Abdulaziz City for Science and Technology (KACST) in partnership with UAVOS. It is the largest of the Saker family of prototype UAVs, which range from Saker-1 to Saker-4 and have their roots in the early 2010s. KACST unveiled an initial mock-up of the Saker-1 in 2017, announcing that the aircraft was capable of equipping with bombs and missiles. A promotional video released by the company showed personnel attaching Chinese-made AR-1 guided missiles and FT-9 guided bombs to the Saker-1. In March 2020, KACST unveiled the upgraded Saker-1B, which has an improved endurance over the Saker-1 and is capable of autonomous take-off and landing, according to the manufacturer. In an interview with Janes, UAVOS CEO Aliaksei Stratsilatau said that UAVOS and KACST have produced three Saker-1B prototypes and conducted over 1,000 hours of take-off and landing test missions. In May 2020, KACST and UAVOS announced plans to produce the Saker-1C, which will reportedly have an improved payload capacity, range, and flight ceiling. As of this writing, the Saker-1 remains in development and does not appear to be in active service.

Specifications

	Saker-1B
Class	III
Status	B/B
Unveiled	2017
Operators	
Length	
Width (wingspan)	18.7m (61ft)
Height	
Maximum take-off weight	1,100kg (2,400lb)
Speed (max)	
Range	
Endurance	19 hours
Flight ceiling	5,000m (16,000ft)
Payload	
Armament	AR-1, FT-9

KACST Saker-1. (Georg Mader)

AL ARABIYA ENGLISH, 'Saudi Arabia unveils new strategic drone program 'Saqr 1'', YouTube, 14 May 2017,
https://www.youtube.com/watch?v=GyzA18Z4bzw

HOST, P., 'UAVOS KACST unveil Saker-1B autonomous-capable MALE UAV', *Janes*, 6 March 2020,
https://www.janes.com/defence-news/news-detail/uavos-kacst-unveil-saker-1b-autonomous-capable-male-uav

SHEPHARD NEWS TEAM, 'UAVOS and KACST to develop Saker 1C UAV', *Shephard Media*, 26 May 2020,
https://www.shephardmedia.com/news/uv-online/uavos-and-kacst-develop-saker-1c-uav/

SERBIA

EDePor Strsljen

The Strsljen (Hornet) is a demonstrator Class III rotary-wing UCAV developed by Company Engine Development and Production, or EDePro. A mock-up of the Strsljen was unveiled at the 2017 IDEX defence exhibition. In an interview with Janes, EDePro lead engineer Dr Predrag Matejic said that work on the Strsljen began in 2012 in response to customer requirements. It is powered by a single turbo-shaft Phoenix-180 engine. EDePro displayed another mock-up of the Strsljen at the 2018 and 2019 IDEX defence exhibition. As of 2019, EDePro was conducting tests of the Strsljen. In public statements and displays, EDePro has described the Strsljen as a strike-capable UAV and compatible with either the 45kg (99lb) Spider guided missile or the 11.7kg (26lb) S-8 unguided rocket. It is unclear if it has conducted any weapons trials to date.

Specifications

Class	III
Status	B/C
Unveiled	2017
Operators	
Length	8.8m (29ft)
Width (main rotor diameter)	7.6m (25ft)
Height	2.4m (8ft)
Maximum take-off weight	750kg (1,650lb)
Speed (cruise)	160km/h (86.4kts)
(maximum)	180km/h (97.2kts)
Range	
Endurance	4 hours
Flight ceiling	4,000m (13,000ft)
Payload	350kg (770lb)
Armament	(S-8), (Spider)

SRB Strsljen. (Igor Salinger)

VTI Pegaz

The Pegaz, or Pegasus, is a demonstrator Class II fixed-wing UCAV developed by the Military Technical Institute (VTI). The Pegaz features high-mounted wings, twin-booms, a low-mounted horizontal stabilizer, and a tapered nose. It has fixed landing gear and a push propeller. According to Shephard News, VTI displayed a Pegaz prototype at the Partner defence exhibition in Serbia in 2011, the same year that it reportedly conducted its maiden flight. VTI displayed a Pegaz fitted with unidentified anti-tank missiles at a military parade in Belgrade in 2016. Following Serbia's acquisition of Chinese-made CH-92As in 2020, Serbian media outlets reported that VTI was working with engineers from China on arming the Pegaz with Chinese missiles.

Specifications

Class	II
Status	A/C
Unveiled	2011
Operators	
Length	5.4m (17ft)
Width (wingspan)	6.34m (21ft)
Height	
Maximum take-off weight	230kg (500lb)
Speed (maximum)	210km/h (113.4kts)
Range	50km (31 miles)
Endurance	8 hours
Flight ceiling	3,000m (9,843ft)
Payload	55kg (31lb)
Armament	

Bastiaans, P., 'Serbia exhibits latest UAVs', *Shephard News*, 7 September 2012, https://www.shephardmedia.com/news/uv-online/serbia-exhibits-latest-uavs/

Golavic, M., 'Српски „пегаз' са кинеским ракетама', *Politika.rs*, 25 July 2020, http://www.politika.rs/scc/clanak/459105/Srpski-pegaz-sa-kineskim-raketama

Malyasov, D., 'Serbia unveils new «Pegaz» multifunctional unmanned aircraft complex with anti-tank missiles', Defence Blog (blog), 23 October 2016, https://defence-blog.com/news/serbia-unveils-new-pegaz-multi-functional-unmanned-aircraft-complex-with-anti-tank-missiles.html

VTI Pegaz seen in 2017 with an SGW M 120 light guided missile. (Igor Salinger)

YugoImport Gavran

The Gavran is a family of prototype fixed-wing loitering munitions developed by Yugoimport and comprised of, as of this writing, the Gavran and Gavran-145. The Gavran features a tubular fuselage with high-mounted wings and a twin-boom tail with a high-mounted tailplane. It employs a single two-cylinder engine in the pusher configuration and a solid-propellant booster for launch, which can be mounted on a vehicle. The Gavran-145, meanwhile, features more of a conventional design for loitering munitions, with a slender, tubular fuselage and high-mounted wings that expand upon launch. It employs a single engine in the pusher configuration and is container-launched using a solid fuel booster. The truck-mounted container launch system can accommodate up to 27 containers. According to the manufacturer, Gavran loitering munitions are intended to be deployed in 'swarms of several vehicles' that are controlled by a single operator using radio datalinks. Yugoimport unveiled the Gavran and Gavran-145 in 2021; it is unclear whether it has conducted flight or weapons trials as of this writing.

Specifications

	Gavran	Gavran-145
Class	Loitering munition	Loitering munition
Status	C	C
Unveiled	2021	2021
Operators		
Length	4.5m (15ft)	2.2m (7.2ft)
Width (wingspan)	3.2m (10.5ft)	2.4m (8ft)
Height	0.7m (2ft)	0.4m (1.3ft)
Maximum take-off weight	50kg (110lb)	50kg (110lb)
Speed (max)		150km/h (81kts)
Range	150km (93 miles)	150km (93 miles)
Endurance	0.5 hours	
Flight ceiling	1,500m (4,920ft)	2,000m (6,560ft)
Payload	13kg (29lb)	15kg (33lb)
Armament		

'PRVI SRPSKI DRON UBICA: "Gavran" može da uništi bilo koji cilj, a jedna stvar čini ga savršenim za bojište', *Srbija Danas*, 9 March 2021, https://www.srbijadanas.com/vesti/naoruzanje/srpski-dron-ubica-gavran-srpsko-oruzje-srpsko-naoruzanje-2021-03-08

Tony (@cyberspec1), 'New Gavran-145 (#Raven_145) Kamikaze drone (loitering munition)', Twitter, 29 June 2021, https://twitter.com/Cyberspec1/status/1410061824121208832

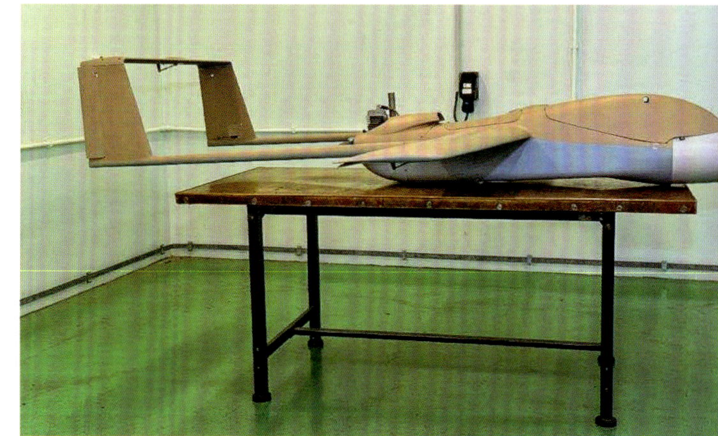

YugoImport Gavran. (YugoImport)

SINGAPORE

ST Engineering Stinger

The Stinger is a Class I rotary-wing UCAV developed by ST Engineering. It is designed to be equipped with a 5.56mm light machine gun and can carry a 100-round ammunition drum. ST Engineering unveiled the STINGER at the 2018 Eurosatory defence exhibition, where it was displayed with the Ultimax 100 light machine gun. According to ST Engineering, the STINGER conducted weapons trials in 2017.

Specifications

Class	I
Status	B/B
Unveiled	2018
Operators	
Length	
Width (wingspan)	
Height	
Maximum take-off weight	
Speed (maximum)	
Range	
Endurance	
Flight ceiling	
Payload	25kg (55lb)
Armament	5.56 light machine gun

SGP Stinger. (ST Engineering)

SHEPHARD NEWS TEAM, 'Singapore Airshow 2018: Soldier system pushes the boundaries', *Shephard Media*, 7 February 2018, https://www.shephardmedia.com/news/landwarfareintl/singapore-airshow-2018-soldier-system-pushes-bound/

SOUTH AFRICA

Denel Dynamics Seeker 400

The Seeker 400, or S400, is a Class II fixed-wing UAV developed by Denel Dynamics. It features high-mounted wings and a twin-boom, pusher-propeller design. The Seeker 400 is an improved version of the Seeker II and part of Denel's Seeker family of UAVs, the first versions of which were fielded by the South African military in the mid-1980s. Denel Dynamics unveiled the Seeker 400 at the 2008 Africa Aerospace and Defence exhibition. The Seeker 400 conducted its maiden flight in 2014. Denel Dynamics describes the Seeker 400 as multi-role ISR system capable of conducting strikes. At the 2010 Africa Aerospace and Defence exhibition, it exhibited a mock-up of the Seeker 400 fitted with Impi laser-guided missiles. At the 2015 IDEX defence exhibition, Denel displayed the Seeker 400 alongside the 15kg (33lb) Impi-S and 45kg (99lb) Mokopa anti-tank missiles. The Seeker 400 may also be compatible with the P2, a 14kg (31lb), GPS-guided anti-tank missile designed for UAVs. According to its 2017/2018 annual report, Denel had received an order from an unnamed customer in the Middle East for a weaponized version of the Seeker 400 that was integrated with the P2. At the UMEX defence exhibitionin February 2020, Denel displayed the Seeker-400 with both the Impi-S and P2 missiles. The Seeker 400 is believed to be in military service with South Africa and the United Arab Emirates, although it is not confirmed whether either country operates the armed version.

Specifications

Class	II
Status	A/A
Unveiled	2008
Operators	South Africa, UAE
Length	
Width (wingspan)	5.7m (18.7ft)
Height	
Maximum take-off weight	550kg (1,210lb)
Speed	
Range (LOS)	250km (155 miles)
Endurance	16 hours
Flight ceiling	
Payload	100kg (220lb)
Armament	Impi–S, P2, (Mokopa)

DEFENCEWEB, 'Weaponised Seeker 400 debuts at IDEX', *DefenceWeb*, 24 February 2015,
https://www.defenceweb.co.za/aerospace/aerospace-aerospace/weaponised-seeker-400-debuts-at-idex/

DEFENCEWEB, 'More details emerge on Seeker 400 export order', *DefenceWeb*, 15 March 2019,
https://www.defenceweb.co.za/aerospace/aerospace-aerospace/more-details-emerge-on-seeker-400-export-order/

DEFENCEWEBTV, 'UMEX 2020 International Unmanned Defence Systems and Training Exhibition Abu Dhabi UAE Day 2', YouTube, 25 February 2020,
https://www.youtube.com/watch?v=DB5OV_h4XN8

LAKE, D., 'AAD 10: Seeker 400 ready in two years', *Shephard News*, 23 September 2010,
https://www.shephardmedia.com/news/uv-online/aad-10-seeker-400-ready-in-two-years/

SHEPHARD NEWS TEAM, 'Denel's Seeker 400 UAV completes first flights', *Shephard News*, 27 February 2014,
https://www.shephardmedia.com/news/uv-online/denels-seeker-400-uav-completes-first-flights/

MILKOR MA380

The MA380 is a demonstrator Class II fixed-wing UAV developed by MILKOR. It features low-mounted wings, retractable landing gear, a T-tail design. With a range of up to 2,000km (1,243 miles) when equipped with a satellite communications capability, MILKOR has described the MA380 as a medium-altitude long-endurance UAV. MILKOR unveiled the MA380, the largest of a family of three UAVs, at the 2018 Defence Services Asia exhibition. According to the South African company, the MA380 can be fitted with hardpoints for weapons, although it did not appear with any in released photographs.

Specifications

Class	II
Status	B/C
Unveiled	2018
Operators	
Length	6.0m (19.7ft)
Width (wingspan)	12.0m (39ft)
Height	
Maximum take-off weight	650kg (1,430lb)
Speed (maximum)	230km/h (124.2kts)
Range (LOS)	250km (155 miles)
(BLOS)	1,000km (620 miles)
Endurance	20 hours
Flight ceiling	5,500m (18,000ft)
Payload	150kg (330lb)
Armament	

MILKOR MA380 at Waterkloof in 2019. (via Georg Mader)

MILKOR, 'DSA:2018 MILKOR Unveils MA 80 380 UAS', news release, April 2018, https://milkor.com/dsa-2018-milkor-unveils-ma-80-380-uas/

MILKOR, 'MA 380', Air Systems, accessed on 15 February 2021, https://milkor.com/ma-380/

A weaponized Seeker 400 – also known as the Snyper – is seen here armed with the Mokopa anti-tank missiles. (DENEL)

MILKOR MA380L

The MILKOR UCAV is a concept Class III fixed-wing UCAV developed by MILKOR UAE, the United Arab Emirates-based office of the South African defence firm MILKOR. It features low-mounted wings and a conventional V-tail vertical stabilizer. MILKOR unveiled a full-scale mock-up of the MILKOR UCAV at the IDEX defence exhibition in February 2021, where it was displayed fitted with multiple Halcon Desert Sting precision guided munitions on two hardpoints under each wing. The UAE's Halcon Systems Desert Sting family of precision guided munitions are advertised with a variety of sizes ranging from 8 to 35kg (18 to 77lb). As of this writing, it is unclear whether the UCAV has conducted flight or weapons trials.

Specifications

Class	III
Status	C/C
Unveiled	2021
Operators	
Length	9.0m (30ft)
Width (wingspan)	18m (59ft)
Height	
Maximum take-off weight	1,300kg (2,860lb)
Speed (cruise)	150km/h (81kts)
(maximum)	250km/h (135kts)
Range (LOS)	350km (217 miles)
(BLOS)	2,000km (1,240 miles)
Endurance	35 hours
Flight ceiling	9,144m (30,000ft)
Payload	210kg (462lb)
Armament	(Desert Sting 16), (Desert Sting-35)

MILKOR, 'Milkor UCAV IDEX', YouTube, 22 February 2021, https://www.youtube.com/watch?v=vHxjv5fO-vQ

MILKOR, 'Milkor UCAV', Air Systems, accessed on 23 February 2021, https://milkor.com/ucav/

SECRET PROJECTS FORUM, 'South African missiles/rockets/PGM's - Prototypes, Projects, Concepts, etc.', 17 September 2018, https://www.secretprojects.co.uk/threads/south-african-missiles-rockets-pgms-prototypes-projects-concepts-etc.21179/page-6

The Milkor MA380L displayed with the Desert Sting 16 and Desert Sting 35 missiles at the IDEX 2021 exhibition. (via Georg Mader)

Paramount N-Raven

The N-Raven is a fixed-wing loitering munition developed by Paramount Group. It was unveiled at the 2021 IDEX defence exhibition. Paramount Group has described the N-Raven as a low-cost 'swarming system' that is designed to saturate enemy defences. It can be launched from ground, air, and maritime platforms. Paramount describes the N-Raven as a 'family' of aircraft, which likely means that it will encompass a variety of mission-oriented systems in the future such as ISR and attack variants.

Specifications

Class	Loitering munition
Status	C
Unveiled	2021
Operators	
Length	
Width (wingspan)	
Height	
Maximum take-off weight	41kg (90lbs)
Speed (cruise)	180km/h (97.2kts)
Range	250km (155 miles)
Endurance	1.5 hours
Flight ceiling	
Payload	15kg (33lb)
Armament	

The Paramount N-Raven. (Paramount)

Helou, A., 'Paramount Group pitches new drone swarm amid region's lack of countermeasures', *DefenseNews*, 22 February 2021, https://www.defencenews.com/digital-show-dailies/idex/2021/02/22/paramount-group-pitches-new-drone-swarm-amid-regions-lack-of-countermeasures/

SOUTH KOREA

KAI Devil Killer

The Devil Killer is a family of fixed-wing loitering munitions developed by Korea Aerospace Industries (KAI) and comprised of the DK-20, DK-25, and DK-150. The basic design features a rectangular fuselage with a tapered nose and high-mounted, swept-back wings that expand upon launch. It employs two rear-mounted electric ducted fan propellers and a canister launch system. KAI unveiled the original, 25kg (55lb) Devil Killer in 2012, which it said it developed in response to a request for proposals from South Korea's Defense Acquisition Program Administration. A Devil Killer prototype reportedly conducted its maiden flight in 2011. At the Dronebot Combat Development Conference in 2018, KAI unveiled the smaller DK-20 and larger DK-150 variants of the Devil Killer. According to media reports, KAI outfitted the DK-150 with an optical infrared sensor to meet a Republic of Korea Army requirement. However, it is unclear whether any version of the Devil Killer is in active service.

Specifications

	DK–20	DK–25	DK–150
Class	Loitering munition	Loitering munition	Loitering munition
Status	B	B	B
Unveiled	2018	2012	2018
Operators			
Length		1.5m (4.9ft)	
Width (wingspan)		1.3m (4.3ft)	
Height		0.27m (0.9ft)	
Maximum take-off weight	20kg (44lb)	25kg (55lb)	150kg (330lb)
Speed (max)	250km/h (135kts)	400km/h (216kts)	150km/h (81kts)
Range	70km (43 miles)	40km (25 miles)	80km (50 miles)
Endurance			
Flight ceiling			
Payload			
Armament			

'Kᴀɪ, 육군의 드론봇 전투체계 위한 최적 솔루션 선봬', *KN News*,
4 April 2018,
http://www.knnews.co.kr/news/articleView.php?idxno=1245454

Wᴀʟᴅʀᴏɴ, G., 'KAI developing 'suicide combat UAV', *FlightGlobal*,
28 September 2012,
https://www.flightglobal.com/kai-developing-suicide-combat-uav/107220.article

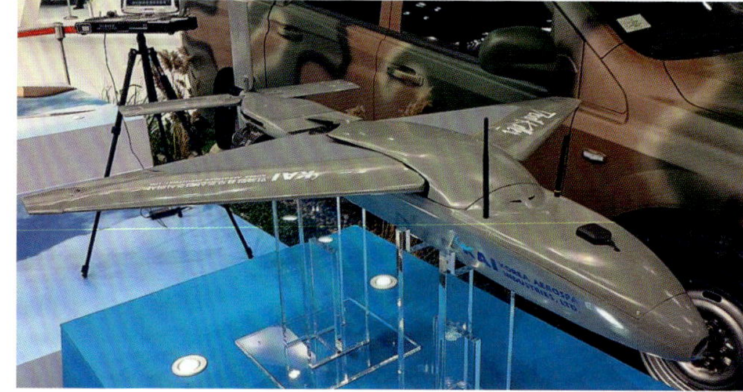

KAI Devil Killer. (via Korean forums)

KAI K-UCAV

The K-UCAV is a conceptual Class III fixed-wing UCAV developed by Korea Aerospace Industries (KAI). Work on the K-UCAV began in 2008 and KAI revealed a subscale model of the aircraft the following year at the 2009 ADEX exhibition in Seoul. The K-UCAV is intended to serve as a stealthy, jet-powered attack platform that could conduct air-to-air and air-to-ground missions deemed too risky for crewed aircraft. The original design for the K-UCAV featured a slender fuselage with a V-tail vertical stabilizer. In 2016, KAI revealed a new, tailless design for the K-UCAV featuring a blended wing-body and called the K-X, a subscale model of which was displayed at the 2017 ADEX exhibition. KAI has conducted flight tests of a subscale version of the K-UCAV and, as of 2019, work on the K-UCAV was ongoing, though the current status of the project remains unclear. KAI and KAL have offered competing designs for the South Korean military's Agency for Defense Development's Kaori-X UCAV project.

Specifications

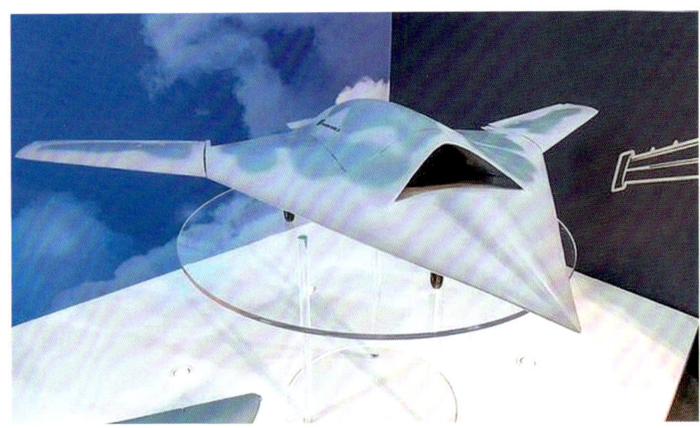

KAI K-UCAV. (Hykonnen)

Class	III
Status	C/C
Unveiled	2009
Operators	
Length	7.6m (25ft)
Width (wingspan)	10.6m (35ft)
Height	3.0m (10ft)
Maximum take-off weight	3,910kg (8,602lb)
Speed (max)	1,173km/h (633.4kts)
Range	280km (174 miles)
Endurance	
Flight ceiling	12,000m (40,000ft)
Payload	
Armament	

ARTHUR, G., 'South Korean firms display new and future UAVs', *Shephard News*, 25 October 2019,
https://www.shephardmedia.com/news/uv-online/south-korean-firms-display-new-and-future-uavs/

GOVINDASAMY, S., 'KAI outlines work on indigenous UCAV design', *FlightGlobal*, 28 October 2009,
https://www.flightglobal.com/kai-outlines-work-on-indigenous-ucav-design/89899.article

KAL KUS-FC

The KUS-FC is a conceptual Class III fixed-wing UCAV developed by Korean Air – Aerospace Business Division (KAL-ASD). In 2011, KAL displayed a subscale mock-up of an initial design for the KUS-FC, which it said could be used as a stealthy attack aircraft. According to KAL, the KUS-FC will feature a tailless blended wing-body design with swept-back wings, a turbojet engine, and an internal weapons bay. Work on the KUS-FC appears to be ongoing, but the status of the project is unclear. As of 2020, KAL was working on a full-scale model of the KUS-FC in preparation for future flight testing but lacked funding for the tests themselves. KAL has offered the KUS-FC as a potential platform for the South Korean military's Agency for Defense Development's Kaori-X UCAV project.

Specifications

Class	III
Status	C/C
Unveiled	2009
Operators	
Length	10m (33ft)
Width (wingspan)	16m (53ft)
Height	3.0m (10ft)
Maximum take-off weight	
Speed (max)	
Range	
Endurance	6 hours
Flight ceiling	
Payload	
Armament	

KAL KUS-FC. (via Georg Mader)

Arthur, G., 'ADEX 2017: Korean Air promotes crewless aircraft', *Shephard News*, 17 October 2017,
https://www.shephardmedia.com/news/uv-online/adex-2017-korean-air-promotes-crewless-aircraft/

Aviation Week Staff, 'Korean Air Preparing Jet Drone Demo Design', *Aviation Week*, 24 February 2020,
https://aviationweek.com/defense-space/korean-air-preparing-jet-drone-demo-design

KAL KUS-FS

The KUS-FS is a prototype Class III fixed-wing UAV developed by Korean Air – Aerospace Business Division (KAL-ASD). Also known as the Medium-Altitude Unmanned Aircraft (MUAV), the KUS-FS features low-mounted wings, a bulbous nose section, and a conventional V-tail with a single downward vertical plane. It employs a single engine – reportedly a Pratt & Whitney PT6 – in the pusher configuration and requires a runway for launch and recovery. KAL displayed a subscale model of the KUS-FS at the ADEX 2017 defence exhibitionand a full-scale prototype at ADEX 2019, though the KUS-FS reportedly conducted its maiden flight as early as 2012. Photos of the KUS-FS show it equipped with two hardpoints under each wing, indicating that is designed to be capable of conducting strikes. An investigation in 2021 by South Korean lawmakers found that the MUAV project is not expected to be completed until March 2022, a delay of some 53 months over the initial timeline. Difficulties with air-to-ground communications and wing icing have contributed to the postponement. Both the Korean Air Force and Army have expressed interest in the acquiring the aircraft.

Specifications

Class	III
Status	B/B
Unveiled	2017
Operators	
Length	13m (43ft)
Width (wingspan)	25m (82ft)
Height	3m (10ft)
Maximum take-off weight	4,100kg (9,020lb)
Speed (max)	
Range	1,850km (1,150 miles)
Endurance	30 hours
Flight ceiling	13,000m (42,560ft)
Payload	
Armament	

KUS-FS prototype. (Hwangbo Joonwoo)

Gordon, A., 'ADEX 2019: Korean MALE UAV prototype breaks cover', *Shephard News*, 18 October 2019,
https://www.shephardmedia.com/news/uv-online/adex-2019-korean-male-uav-prototype-breaks-cover/

Kim, D. Y., 'South Korean MND eyes KUS-FS UAV for RoKA Ground Operations Command', *Janes*, 30 March 2020,
https://www.janes.com/defence-news/news-detail/south-korean-mnd-eyes-kus-fs-uav-for-roka-ground-operations-command

성상훈 기자, '[단독] 5000억 쓴 '대북 무인 정찰기' 사업, 53개월 지연됐다', *Korean Economic Daily*, 27 August 2021,
https://www.hankyung.com/politics/article/202108272550i

KAL KUS-VH

The KUS-VH is a prototype Class III rotary-wing UCAV developed by Korean Air – Aerospace Business Division (KAL-ASD). This UCAV is an uncrewed version of the MD Helicopters MD500 Little Bird light attack and observation helicopter, which South Korea licensed for domestic production in the 1970s. In a push to repurpose aging MD500s as uninhabited combat aircraft, KAL began working on the KUS-VH in 2014. KAL unveiled a full-size mock-up of the aircraft at the 2015 ADEX defence exhibition, where it was displayed outfitted with unspecified air-to-ground munitions and rockets. In July 2019, KAL announced that a fully uncrewed KUS-VH had conducted its first flight at a facility in Goheung. KAL displayed the KUS-VH at the 2019 ADEX exhibition fitted with unspecified munitions. According to press reports, the KUS-VH could be equipped with 70mm (2.75in) rockets and air-to-surface missiles, though it is unclear whether weapons trials have occurred as of this writing.

Specifications

Class	III
Status	B/C
Unveiled	2015
Operators	
Length	9.4m (30.8ft)
Width (wingspan)	8.0m (26.2ft)
Height	3.0m (9.8ft)
Maximum take-off weight	4,100kg (9,020lb)
Speed (max)	
Range	
Endurance	6 hours
Flight ceiling	
Payload	440kg (968lb)
Armament	

KUS-VH. (KAL)

CHUANREN, C., 'Unmanned Korean MD500 flies for the first time', *Aviation International Online*, 2 August 2019,
https://www.ainonline.com/aviation-news/defense/2019-08-02/unmanned-korean-md500-flies-first-time

WALDRON, G., 'Pictures: KAL-ASD to test unmanned MD500', *Flight Global*, 20 October 2015,
https://www.flightglobal.com/pictures-kal-asd-to-test-unmanned-md500/118587.article

SPAIN

Aertec Solutions Tarsis 25 and Taris 75

The Tarsis 25 and Tarsis 75 are two demonstrator Class I fixed-wing UCAVs developed by Aertec Solutions. With a 75kg (165lb) maximum take-off weight, the Tarsis 75 is the larger of the two aircraft and features mid-mounted wings and a mid-mounted horizontal stabilizer. The Tarsis 25 is smaller and has high-mounted wings and a low-mounted horizontal stabilizer. Both aircraft feature twin booms to swept-back fins, fixed landing gear, and a pusher propeller. Aertec unveiled the Tarsis platform at the 2015 Paris Air Show, which it displayed with a model of the 5kg (11lb) Fox air-to-ground guided rocket, also known as the Light Aerial Platform Micro-Missile. Although Aertec markets the Tarsis aircraft as primarily intended for intelligence-gathering, it has displayed the Tarsis alongside this semi-active laser-guided lightweight rocket at defence exhibitions such as Homesec 2017, among others. At the Expodefence 2019 defence exhibition, an Aertec official told *Janes* that both the Tarsis 25 and Tarsis 75 could perform a strike mission. In 2020, Spain's Ministry of Defence awarded Aertec a contract to develop a lightweight guided weapon for a Class I UAV. In January 2021, the Spanish Army completed an evaluation of the unarmed Tarsis 75 for artillery fire correction and ISR as part of Spain's Rapaz Programme, though it does not appear to be in active service with the Spanish military. At the IDEX defence exhibition in February 2021, Aertec displayed the Tarsis 75 with one Fox guided missile fitted under each wing.

Specifications

	Tarsis 25	Tarsis 75
Class	I	I
Status	B/C	B/C
Unveiled	2015	2015
Operators		
Length	2.5m (8.2ft)	3.8m (12.5ft)
Width (wingspan)	4.0m (13ft)	5.2m (17ft)
Height	0.9m (2.9ft)	1.0m (3.1ft)
Maximum t/o weight	25kg (55lb)	75kg (165lb)
Speed (cruise)	100km/h (54kts)	100km/h (54kts)
Range (operational)	150km (93 miles)	150km (93 miles)
Endurance	8 hours	12 hours
Flight ceiling	4,000m (13,120ft)	5,000m (16,400ft)
Payload	5.0kg (11lb)	12kg (26lb)
Armament	(A-Fox)	(A-Fox)

Tarsis 25 (front) and Tarsis 75 (background). (Aertec)

Defensa.com, 'El RPAS Tarsis 75 de Aertec completa la ultima evaluación operativa en el marco del Programmea Rapaz del Ministerio de Defensa', *Defensa.com*, 27 January 2021, https://www.defensa.com/espana/rpas-tarsis-75-aertec-completa-ultima-evaluacion-operativa-marco

Garcia, D., 'Aertec presenta en Oriente Próximo su RPAS Tasis 75 armado con cohetes guiados Fox', *Infodron.es*, 28 February 2021, http://www.infodron.es/id/2021/02/28/noticia-aertec-presenta-oriente-proximo-tarsis-armado-cohetes-guiados.html

Host, P., 'Expodefence 2019: Aertec Solutions offers armament-capable Tarsis 25 UAV', *Janes*, 3 December 2019, https://www.janes.com/defence-news/news-detail/expodefensa-2019-aertec-solutions-offers-armament-capable-tarsis-25-uav/

Navarro, J. M., 'Defensa estudia de nuevo armar sus UAVs', *Defensa.com*, 14 February 2020, https://www.defensa.com/espana/defensa-estudia-nuevo-armar-uavs

Thomas, R., 'Paris Air Show: Aertec offers platform and payload', *Shephard News*, 18 June 2015, https://www.shephardmedia.com/news/uv-online/paris-air-show-aertec-offers-platform-and-payload/

TAIWAN

NCSIST Chien Hsiang

The Chien Hsiang (劍翔) is a fixed-wing loitering munition developed by the National Chung-Shan Institute of Science and Technology (NCSIST). The Chien Hsiang bears a strong physical resemblance to the IAI Harpy. It features a delta-wing design with a tubular fuselage and an electric motor in the pusher configuration. According to media reports, the Chien Hsiang is designed to suppress adversary air defences, particularly radar-emitting weapons systems such as radars. NCSIST publicly unveiled the Chien Hsiang in 2017, though it may have made its maiden flight as early as 2011. At the 2019 Taipei Aerospace and Defense Technology Exhibition, NCSIST displayed the Chien Hsiang alongside a trailer-based canister launch system that can accommodate 12 loitering munitions. Taiwan's Ministry of National Defense has reportedly ordered 104 Chien Hsaing systems, which are expected to be delivered between 2019 and 2025 and fielded by the ROC Air Force.

Specifications

Class	Loitering munition
Status	A
Unveiled	2017
Operator	Taiwan
Length	
Width (wingspan)	
Height	
Maximum take-off weight	
Speed (maximum)	
Range	
Endurance	
Flight ceiling	
Payload	
Armament	

The Chien Hsiang employs a trailer-based canister launch system. (NCSIST)

Au, C., 'TADTE 2019: Taiwan's 'Harpy' enters series production', *Shephard News*, 15 August 2019, https://www.shephardmedia.com/news/uv-online/tadte-2019-taiwans-harpy-enters-series-production/

Wong, K., 'TADTE 2019: Taiwan's NCSIST rolls out indigenous anti-radiation loitering munition', *Janes*, 19 August 2019, https://www.janes.com/defence-news/news-detail/tadte-2019-taiwans-ncsist-rolls-out-indigenous-anti-radiation-loitering-munition

NCSIST Fire Cardinal

The Fire Cardinal (火紅雀) is a fixed-wing loitering munition developed by the National Chung-Shan Institute of Science and Technology (NCSIST). It is a strike variant of the Cardinal, a Class I fixed-wing reconnaissance UAV. NCSIST unveiled the Fire Cardinal at the 2019 Taipei Aerospace and Defense Technology Exhibition. Unlike the Cardinal, which is powered by a tractor configuration, the Fire Cardinal has two wing-mounted electric motors. According to NCSIST, the Fire Cardinal is embedded with an 'intelligence object detection system' that is uses artificial intelligence and machine learning to detect and track targets. The Fire Cardinal remains under development.

Specifications

Class	Loitering munition
Status	B
Unveiled	2019
Operators	
Length	1.2m (3.9ft)
Width (wingspan)	2.0m (6.6ft)
Height	0.6m (2.0ft)
Maximum take-off weight	6.0kg (13lb)
Speed (cruise)	
Range	
Endurance	
Flight ceiling	
Payload	
Armament	

The NCSIST Fire Cardinal on display in 2019. (via UAS Vision)

Wong, K., 'TADTE 2019: NCSIST unveils Fire Cardinal mini-UAV, *Janes*, 16 August 2019,
https://www.janes.com/defence-news/news-detail/tadte-2019-ncsist-unveils-fire-cardinal-mini-uav

'全面攻佔真實版！國產自主武器「火紅雀」！', *UMedia*, 22 August 2019,
https://www.umedia.world/news_details.php?n=2019082213350172583

NCSIST Teng Yun

The Teng Yun (騰雲) is a prototype Class III fixed-wing UCAV developed by the National Chung-Shan Institute of Science and Technology (NCSIST). It features a conventional design for medium-altitude long-endurance UAVs, with a slender fuselage and bulging nosecone, mid-mounted wings, and a V-tail with a single downward vertical plane. It employs a single engine in the pusher configuration and requires a runway for take-off and landing. Work on the Teng Yun began in 2009 and NCSIST displayed subscale models of a possible MALE UAV design at the 2011 Taipei International Aerospace and Defense Industry Exhibition (TADTE). NCSIST unveiled a full-scale mockup of the Teng Yun at the 2015 TADTE defence exhibition. At the time, NCSIST representatives said that the Teng Yun would be armed with air-to-surface missiles. At the 2017 IDEX defence exhibition, NCSIST displayed a subscale model of the Teng Yun equipped with Hellfire-like missiles on one hardpoint under each wing. The Teng Yun conducted its maiden flight in 2018. Media reports in 2020 indicated that the Teng Yun was undergoing weapons trials, though it is unclear which rockets and missiles will comprise the Teng Yun's eventual loadout. NCSIST has produced at least four Teng Yun demonstrator aircraft in two configurations. While the initial Teng Yun configuration featured a Rotax 914F engine, NCSIST replaced it with a Honeywell TPE331 turboprop in the second configuration based on feedback from the Republic of China Air Force. The second iteration also features a redesigned fuselage and larger wings. The Teng Yun is believed to remain under development.

Specifications

Class	III
Status	B/B
Unveiled	2015
Operators	
Length	8.0m (26ft)
Width (wingspan)	18m (59ft)
Height	
Maximum take-off weight	
Speed (max)	
Range	1,000km (620 miles)
Endurance	24 hours
Flight ceiling	7,620m (25,000ft)
Payload	
Armament	

Teng Yun. (Liberty Times/You Tai-lang)

Cole, J. M., 'Taiwan Unveils New Long-Endurance Drone, New Weapons at Defense Trade Show', *The Diplomat*, 13 August 2015, https://thediplomat.com/2015/08/taiwan-unveils-new-long-endurance-drone-new-weapons-at-defense-trade-show/

'Taiwan displays armed MALE UAV at IDEX', *Janes*, 1 March 2017, http://www.janes.com/article/68384/taiwan-displays-armed-male-uav-at-idex

'國防MIT二代騰雲機傳掛載劍一飛彈測試 驗證偵察及攻擊戰力', *Liberty Times Net*, 13 November 2020, https://news.ltn.com.tw/news/politics/breakingnews/3350568

THAILAND

RV Connex Sky Scout/Sky Scout-X

The Sky Scout-X is conceptual Class II fixed-wing UAV developed by RV Connex. It is an armed variant of the Sky Scout, an unarmed UAV for intelligence-gathering that is already in already in active military service with the Royal Thai Air Force (RTAF). The Sky Scout-X features high-mounted wings and a twin-boom design with a high-mounted tailplane. RV Connex displayed the Sky Scout-X armed with two Thales Freefall Lightweight Multirole Missiles (FFLMM) at the 2019 Defense and Security exhibition, where it announced that the company was working with the RTAF to develop an operational variant of the aircraft. The following year, the RTAF displayed the RTAF U1-M at the 2020 Singapore Airshow. The RTAF U1-M is believed to be either in service with or on order for the Royal Thai Air Force.

Specifications

	SkyScout
Class	II
Status	A/A
Unveiled	2019
Operators	
Length	3.6m (11.8ft)
Width (wingspan)	6.2m (20.3ft)
Height	
Maximum take-off weight	140kg (308lb)
Speed (maximum)	107km/h (57.7kts)
Range	100km (62 miles)
Endurance	6 hours
Flight ceiling	
Payload	10kg (22lb)
Armament	(FFLMM)

Sky Scout armed with two Thales Lightweight Multirole Missiles. (RTAF)

ARTHUR, G., 'D&S 2019: Thai military charts course towards armed UAVs', *Shephard News*, 18 November 2019, https://www.shephardmedia.com/news/uv-online/ds-2019-thai-military-charts-course-towards-armed-/

DONALD, D., 'Thai UAV Surprises at Singapore Show', *Aviation International Online*, 12 February 2020, https://www.ainonline.com/aviation-news/defence/2020-02-12/thai-uav-surprises-singapore-show

GREVATT, J., 'D&S 2019: RV Connex builds Thai profile', *Janes*, 22 November 2019, https://www.janes.com/defence-news/news-detail/ds-2019-rv-connex-builds-thai-profile

TURKEY

Asisguard Songar

The Songar is a Class I rotary-wing UCAV developed by Asisguard, a division of Asis. Asisguard unveiled the Songar at the IDEF 2019 defence trade show. The Songar is equipped with what Asisguard calls a 'Drone Mounted Machine Gun', or DMMG, that is compatible with 5.56 x 45-millimetre NATO-class ammunition. It has a maximum take-off weight of 45kg (99lb) and an operational range of 3km (1.8 miles). In a 2019 interview with Shephard News, Kadir Kugu, chief of product management at Asis, said that the Songar was developed in response to the Turkish government's specific requirements. In February 2020, Asisguard revealed that it delivered the first Songar systems to the Turkish military for evaluation. In February 2021, Asisguard announced that it had successfully equipped the Songar with the Akdaş AK-40GL grenade launcher.

Specifications

Class	I
Status	A/A
Unveiled	2019
Operators	Turkey
Length	
Width	1.5m (4.8ft)
Height	0.8m (2.3ft)
Maximum take-off weight	45kg (99lb)
Speed (cruise)	36km/h (19.4kts)
Range (operational)	3.0km (2 miles)
Endurance	0.45 hours
Flight ceiling	3,000m (9,840ft)
Payload	9kg (20lb)
Armament	Light machine gun, grenades

Songar. (Asisguard)

Asis Guard, 'Songar Armed Drone', accessed on 1 March 2021,
https://www.asisguard.com/en/project/songar/

CNN Turk staff, 'Türkiye'nin ilk milli bomba atarlı dronu Songar', *CNN Turk*, 28 February 2021,
https://www.cnnturk.com/ekonomi/turkiyenin-ilk-milli-bomba-atarli-dronu-songar?page=1/

Martyr, K., 'IDEF 2019: Asis showcases new spy and weaponised drones', *Shephard News*, 3 May 2019,
https://www.shephardmedia.com/news/uv-online/idef-2019-asis-showcases-new-spy-and-weaponised-dr/

Yıldırım, G., 'Türk Silahlı Kuvvetlerine ilk silahlı drone teslimatı', *Anadolu Ajansi*, 1 February 2020,
https://www.aa.com.tr/tr/bilim-teknoloji/turk-silahli-kuvvetlerine-ilk-silahli-drone-teslimati-/1720977

Baykar Akinci

The Akinci is a demonstrator Class III fixed-wing UCAV developed by Baykar Makina. The Akinci is notable for its gull wing and twin-engine design. The initial demonstrator aircraft was powered by two Ivchenko-Progress AI-450S engines, a Ukrainian-made 750 horsepower turboprop engine. Baykar is working on integrating the Turkish-made Tusaş Engine Industries PD-222, a 250-hoursepower variant of the TEI PD-170 engine. Baykar unveiled the Akinci in mid-2018. The Akinci conducted its first flight in December 2019 at Çorlu Airport, a Turkish Army airfield in north-western Turkey that serves as the test site for the Akinci. Baykar unveiled a second demonstrator aircraft in August 2020 and a third in January 2021. The Baykar website describes the Akinci as capable of equipping with a variety of munitions. In addition to the MAM-L, MAM-C, CIRIT, and L-UMTAS, Baykar claims the Akinci could carry the Mk 81/82/83 guided bombs, the Aselsan Mk-83 wing-assisted guided bomb, the Tubitak-Sage Gokdogan and Bozdogan air-to-air missiles, and the SOM-A cruise missile. In April 2021, an Akinci conducted a successful live-fire test of the Roketsan MAM-T, a new long-range member of the MAM-series guided missiles. A photo published by Baykar in July 2021 showed an Akinci equipped with a Tubitak Sage HGK-84 guided bomb underneath the main fuselage. However, given that the Akinci continues to conduct flight trials and tests, the final loadout of compatible munitions could change. In a ceremony in August 2021, Turkish President Recep Erdogan inducted the first Akinci aircraft into military service.

Specifications

Class	III
Status	A/A
Unveiled	2018
Operators	Turkey
Length	12.2m (40ft)
Width (wingspan)	20.0m (65ft)
Height	4.1m (13ft)
Maximum take-off weight	5,500kg (12,100lb)
Speed (maximum)	360km/h (194.4kts)
Range	
Endurance	24 hours
Flight ceiling	12,192m (40,000ft)
Payload	750kg (1,650lbs)
Armament	HGK-84, MAM-T, (Bozdogan, Bozok, CIRIT, Gokdogan, L-UMTAS, MAM-C, MAM-L, Mk 81/82/83 JDAM, SOM-A)

'AKINCI-C İHA yerli motor PD-222 ile uçacak', *Aydinlik*, 23 January 2021, https://www.aydinlik.com.tr/haber/akinci-c-iha-yerli-motor-pd-222-ile-ucacak-230300

Baykar Defence, 'Bayraktar AKINCI System', accessed on 2 February 2021, https://baykardefence.com/uav-14.html

'Fuat Oktay: Artık ambargolarla yıldırabilecekleri bir Türkiye yok', *Hurriyet*, 6 November 2020, https://www.hurriyet.com.tr/gundem/fuat-oktay-artik-ambargolarla-yildirabilecekleri-bir-turkiye-yok-41656366

Lake, J., 'Turkey's New Raider Takes to the Air', *Aviation International Online*, 11 December 2019, https://www.ainonline.com/aviation-news/defence/2019-12-11/turkeys-new-raider-takes-air

Shephard News Team, 'Akinci UCAV programme progresses with latest tests', *Shephard News*, 7 September 2020, https://www.shephardmedia.com/news/uv-online/akinci-ucav-programme-progresses-latest-tests/

Yaylali, C. D., 'Turkey takes delivery of Akinci armed reconnaissance UAVs', *Janes*, 1 September 2021, https://www.janes.com/defence-news/news-detail/turkey-takes-delivery-of-akinci-armed-reconnaissance-uavs

A Baykar Akıncı on display at Teknofest2021. (Stijn Mitzer and Joost Oliemans)

Bayraktar TB2 armed with four MAM-L munitions. (Baykar)

Baykar Defence Bayraktar-TB2

The Bayraktar-TB2 is a Class III fixed-wing UCAV developed by Baykar Defence. It features a twin-boom, push-propeller design with an inverted V-tail and mid-mounted wings. Work on the Bayraktar-TB2 began in 2007, when the Presidency of Defence Industries (SSM), Turkey's defence armaments directorate, selected Baykar Defence for its new Tactical UAV Project. In 2009, Baykar Defence conducted the first flight of the Bayraktar-Çaldıran, a prototype tactical UAV that was slightly smaller and around 200kg (441lb) lighter than the Bayraktar-TB2. Baykar Defence conducted the first flight tests of the Bayraktar-TB2 in 2014 and displayed a subscale model of the TB2 at the Dimdex 2014 defence exhibition. In December 2015, the Baytakr-TB2 conducted its first weapons trials, launching the Roketsan 37.5kg (83lb) UMTAS anti-tank missile. The following year, in April 2016, the Bayraktar-TB2 conducted a live fire test with the Roketsan 22kg (49lb) MAM-L smart micro munition. In addition to the UMTAS and MAM-L, the Bayratkar-TB2 is also compatible with the 6.5kg (14lb) MAM-C. The Bayraktar-TB2 can carry up to two MAM-L and two MAM-C missiles. In October 2020, CTech Information Technology, a Turkish software company, announced that it had begun deliveries of Satellite Communications On-The-Move terminals to the Turkish military, potentially extending the operational range of the Bayraktar-TB2. In November 2020, a tweet by Baykar Defence suggested that it was in the process of testing satellite-enabled Bayraktar-TB2s.

The Bayraktar-TB2 is in active military service with Azerbaijan, Qatar, Turkey and Ukraine. It is also in service with the Libya Government of National Accord, the U.N.-recognized and Turkish-backed government in Tripoli. Within Turkey, the Bayraktar-TB2 is operated by the Turkish Air Force, Army, Gendarmerie, and General Directorate of Security. With the possible exception of Qatar and the Turkish civilian police agencies, armed versions of the Bayraktar-TB2 are believed to be in service with all domestic and international customers.

Specifications

Class	III
Status	A/A
Unveiled	2014
Operators	Azerbaijan, Libya (GNA), (Poland), Qatar, Turkey, Ukraine
Length	6.5m (21ft)
Width (wingspan)	12m (34ft)
Height	
Maximum take-off weight	650kg (1,430lb)
Speed (maximum)	250km/h (135kts)
Range (LOS)	150km (93 miles)
Endurance	24 hours
Flight ceiling	7,620m (25,000ft)
Payload	155kg (341lb)
Armament	MAM-C, MAM-L, UMTAS

Baykar (@Baykar_Savunma), 'Can Azerbaycan; zaferin kutlu olsun! Eşq olsun Azərbaycan Ordusuna! Qarabağ Azərbaycandır!', Twitter, 10 November 2020, https://twitter.com/Baykar_Savunma/status/1326081511255388160

Baykar Defence, 'Baykar Makina Basın Açıklaması / Bayraktar-Çaldıran Taktik İHA Uçuşları Başarıyla Tamamlandı', news release, 3 October 2009, https://www.baykarsavunma.com/haber-Baykar-Makina-Basin-Aciklamasi---Bayraktar-Caldiran-Taktik-IHA-Ucuslari-Basariyla-Tamamlandi.html

Baykar Defence, 'Taktik İHA Uçuş Testleri Basın Açıklaması', news release, 3 May 2014, https://www.baykarsavunma.com/haber-Tak-tik-IHA-Ucus-Tesleri-Basin-Aciklamasi.html/

Baykar Defence, 'Bayraktar-TB2',, accessed on 14 March 2021, https://baykardefence.com/uav-15.html

Baykar Technologies, 'Bayraktar TB2 Harp Başlıklı MAM-L Atış', YouTube, 29 April 2016, https://www.youtube.com/watch?v=Xx3MEco1Qlw

c4isrnet.com, https://www.c4isrnet.com/artificial-intelli-gence/2021/03/01/turkeys-baykar-begins-designing-ai-powered-combat-drone/?utm_source=twitter.com&utm_medium=social&utm_campaign=Socialflow+DFN

Haber7, 'Milli İHA'ya yerli füze takıldı!', *Haber7*, 18 December 2015, https://www.haber7.com/guncel/haber/1708923-milli-ihaya-yerli-fuze-takildi

'Turkey develops domestic satellite terminals for drones', *Daily Sabah*, 9 October 2020, https://www.dailysabah.com/business/defence/turkey-develops-domestic-satellite-terminals-for-drones

Dasal Dogan

The Dogan is a prototype Class I rotary-wing UCAV developed by Dasal Aviation Technologies, a subsidiary of Aselsan and Altinay. The Dogan features a quadcopter design with eight coaxial motors. It is equipped with a 5.56mm automatic rifle mounted underneath the fuselage and is capable of carrying between 100 and 200 rounds of ammunition. According to the manufacturer, the Dogan has also integrated what Dasal calls an artificial intelligence system on the Dogan that enables it to rapidly identify, classify and track targets. It was unveiled at the IDEF 2021 defence exhibition and has reportedly undergone flight and weapons trials.

Specifications

Class	I
Status	B/B
Unveiled	2021
Operators	
Length	
Width (wingspan)	
Height	
Maximum take-off weight	
Speed (max)	
Range	10km (6 miles)
Endurance	0.5 hours
Flight ceiling	
Payload	
Armament	Assault rifle

The Dasal Dogan. (Dasal Aviation Technologies)

Daily Sabah Staff, 'Turkey's automatic rifle equipped drone almost ready for inventory', *Daily Sabah*, 26 August 2021,
https://www.dailysabah.com/business/defense/turkeys-automatic-rifle-equipped-drone-almost-ready-for-inventory

STM Alpagu

The Alpagu is a prototype fixed-wing loitering munition developed by STM Engineering. Like other fixed-wing loitering munitions such as the Switchblade, the Alpagu features a tubular fuselage with foldable, low-mounted tandem wings and twin vertical stabilizers. The Alpagu is tube-launched and equipped with a 0.3kg (0.66lb) warhead that was developed in partnership with the Turkish firm MKE. Like the Kargu, STM has advertised the Alpagu as capable of conducting operations autonomously. In June 2021, the Turkish defence industries agency (SSB) announced that the Alpagu had successfully completed weapons trials at the Aksaray range. According to media reports, STM intends to produce four versions of the Alpagu for the Turkish military with variations in size and payload capacity. The Alpagu may also be integrated with fixed-wing UCAVs such as the Aksungur or Anka. As of this writing, however, the Alpagu does not appear to be in active service.

Specifications

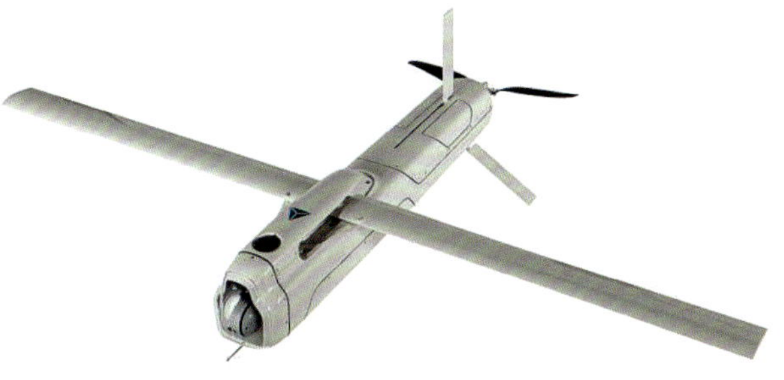

STM Alpagu. (STM)

Class	Loitering munition
Status	B
Unveiled	2017
Operators	
Length	
Width (wingspan)	
Height	
Maximum take-off weight	1.9kg (4lb)
Speed (max)	120km/h (64.8kts)
Range	10km (6 miles)
Endurance	0.25 hours
Flight ceiling	
Payload	0.3kg (1lb)
Armament	

OZBERK, T., 'Alpagu kamikaze hits the target', *Shephard Media*, 30 June 2021,
https://www.shephardmedia.com/news/air-warfare/alpagu-kamikaze-uav-hits-target/

ROJKES DOMBE, A., 'Turkey Unveils Alpagu 2 Tactical Attack UAV', *Israel Defense*, 5 December 2017,
https://www.israeldefense.co.il/en/node/32067?__cf_chl_managed_tk__=pmd_pmNsF_quW22ZQeR9vzol3ND2XFj07UMe62w2VE45eC8-1630168300-0-gqNtZGzNArujcnBszQg9

STM Boyga

The Boyga is a prototype Class I rotary-wing UCAV developed by STM Engineering. It features a quadcopter design with eight coaxial motors. Also referred to as the Ammunition Drop UAV, the Boyga is intended to serve as a platform for 60mm and 81mm mortar rounds. According to STM, the Boyga uses an onboard algorithm to estimate where it should release the mortar rounds over a particular target. STM unveiled the Boyga at the IDEF 2021 defence exhibition.

Specifications

Class	I
Status	C/C
Unveiled	2021
Operators	
Length	0.8m (2.6ft)
Width (wingspan)	0.8m (2.6ft)
Height	0.5m (2ft)
Maximum take-off weight	18kg (40lb)
Speed (max)	
Range	10km (6 miles)
Endurance	0.5 hours
Flight ceiling	2,000m (6,560ft)
Payload	6kg (13lb)
Armament	Mortars

STM Boyga. (STM)

STM Kargu

The Kargu is a rotary-wing loitering munition developed by STM Engineering. The Kargu features a standard micro-rotor design with a rectangular fuselage and four foldable arms. The Kargu-2, an upgraded version of the Kargu, can be equipped with fragmentation, thermobaric, or shaped-charge warheads weighing up to three pounds. STM developed the Kargu's explosive payload in partnership with the Turkish munitions firm MKE. STM unveiled the Kargu in 2017 and, in the following year, announced that it had completed weapons trials. In 2019, the Turkish defence industries agency (SSB) announced that it would acquire 365 Kargu-2s. In advertising materials, STM Engineering has claimed that the Kargu-2 is equipped with facial recognition technology and is capable of conducting some missions autonomously. The Kargu has served as the primary system in STM's KERKES project, which seeks to develop a technical framework for swarming drone operations. STM's claims regarding the Kargu's autonomous capabilities became the focus of intense international attention after a United Nations report in March 2021 suggested that a Kargu-2 may have attacked targets in Libya without human intervention during clashes between Libya's Government of National Accord and Haftar Affiliated Forces. In a June 2021 interview with Nikkei Asia, STM CEO Hakan Guleryuz denied that the Kargu-2 was capable of autonomously engaging targets, claiming that the Kargu's autonomous capabilities were restricted to navigation modes. In addition to Libya, the Kargu has reportedly been used by Turkish or partner forces in Azerbaijan and Syria. In July 2021, STM announced that the first formal export agreement had been concluded with an unidentified country.

Specifications

Class	Loitering munition
Status	A
Unveiled	2017
Operators	Turkey
Length	0.6m (2ft)
Width (wingspan)	0.6m (2ft)
Height	0.4m (1ft)
Maximum take-off weight	7kg (15lb)
Speed (max)	72km/h (38.8kts)
Range	10km (6 miles)
Endurance	0.25 hours
Flight ceiling	2,500m (8,200ft)
Payload	1.3kg (3lb)
Armament	

STM Kargu. (STM)

Shephard News Team, 'STM's Kargu passes precision strike test', *Shephard Media*, 19 July 2021, https://www.shephardmedia.com/news/uv-online/stms-kargu-uas-passes-precision-strike-test/

Tavsan, S., 'Turkish defense company says drone unable to go rogue in Libya', *Nikkei Asia*, 20 June 2021, https://asia.nikkei.com/Business/Aerospace-Defense/Turkish-defense-company-says-drone-unable-to-go-rogue-in-Libya

Yildiz Unal, A., 'Turkey does first exports of 'kamikaze drone' Kargu', *Anadolu Agency*, 21 July 2021, https://www.aa.com.tr/en/economy/turkey-does-first-exports-of-kamikaze-drone-kargu/2310874

TAI Aksungur

The Aksungur, or Anka-2, is a demonstrator Class III fixed-wing UCAV developed by Turkish Aerospace Industries (TAI). The Aksungur features a twin-book design with high-mounted wings. It is powered by two Tusaş Engine Industries PD-170 Dual Turbo Diesel engines, one mounted under each wing. TAI displayed the Aksungur at the IDEF 2019 defence exhibitionin Istanbul in May 2019. It has marketed four variants of the Aksungur: basic, ground attack, SIGINT, and maritime patrol. The basic and SIGINT variants have an endurance of 24 hours, while the ground attack and maritime patrol variants can fly for 12 hours when fully loaded. The ground attack variant features six hardpoints – three under each wing. TAI has proposed equipping the Aksungur with the Roketsan TEBER-81 and TEBER-82 laser-guided bombs, in addition to the suite of CIRIT, MAM-C, MAM-L, and UMTAS munitions that are already compatible with the Anka. At IDEF 2019, TAI also displayed the Aksungur with a mock sonobuoy pod, the ASELSAN miniature bomb, and the Tubitak Sage HGK-82. In September 2020, the SSB, Turkey's armaments ministry, released a video of the Aksungur conducting a weapons trial with a TEBER guided bomb. Also in September, TAI announced that the Aksungur had conducted a 28-hour-flight equipped with a payload of 12 MAM-L guided missiles, two of which were loaded on each of the six hardpoints.

Specifications

Class	III
Status	B/B
Unveiled	2019
Operators	
Length	11.6m (38ft)
Width (wingspan)	24.0m (78ft)
Height	
Maximum take-off weight	3,300kg (7,260lb)
Speed (cruise)	
Range	250km (155 miles)
Endurance	24 hours
Flight ceiling	12,195m (40,000ft)
Payload	750kg (1,650lb)
Armament	TEBER-81/-82, (CIRIT), (HGK-82), (MAM-C), (MAM-L)

An Aksungur fitted with MAM-L guided missiles and TEBER-82 guided bombs at Teknofest 2021. (Cem Dogut)

Turkish Aerospace, 'Aksungur', UAV, accessed on 2 February 2021,
    https://www.tusas.com/en/product/aksungur-unmanned-aerial-vehicle

'Turkish UAV Aksungur flies 28 hours with 12 smart ammunition payloads', *Daily Sabah*, 18 September 2020,
    https://www.dailysabah.com/business/defence/turkish-uav-aksungur-flies-28-hours-with-12-smart-ammunition-payloads

Turnbull, G., 'IDEF: Aksungur UAV makes show debut after first flight', *FlightGlobal*, 2 May 2019,
    https://www.flightglobal.com/military-uavs/idef-aksungur-uav-makes-show-debut-after-first-flight/132517.article

TAI Anka

The Anka is a Class III fixed-wing UCAV developed by Turkish Aerospace Industries (TAI). In 2004, the Turkish Presidency of Defence Industries (SSM) awarded TAI a contract to develop a medium-altitude long-endurance unmanned aerial vehicle. TAI unveiled the Anka at the UK's 2010 Farnborough Air Show. The Anka Block-A, or Anka-A, conducted its first flight in December 2010. In 2012, the SSM announced plans to acquire an initial set of 10 Ankas. In January 2013, the Anka-A reached initial operational capability. Two Anka-As were delivered to the Turkish Air Force in mid-2013 for additional flight tests, one of which crashed in December of that year.

Turkish Aerospace Industries has produced at least three variants of the Anka. The Anka Block-B, or Anka-B, is around 100kg (220lb) lighter than the Anka-A and contains upgraded sensor payloads. It conducted its first flight in January 2015. The Anka-B is equipped with the L3 Wescam CMX-15D FLIR EO/IR and the Aselsan SARPER synthetic aperture radar. With a slightly longer wingspan and fuselage, the Anka-B supplanted the Anka-A as the base Anka platform. The Anka-S is a satellite-enabled variant of the Anka. Equipped with the ViaSat VR-18C Highpower antenna, the Anka-S can conduct beyond line-of-sight operations, extending the Anka-S range far beyond the 250km (155-mile) LOS radius. The Anka-I is an unarmed variant of the Anka designed to gather signals intelligence. The SSM published a photo of the Anka-I in March 2018.

TAI conducted initial weapons trials with the Anka-A and the Roketsan CIRIT 2.75in semi-active laser-guided air-to-ground missiles in May 2013. The Anka-B and Anka-S can carry the CIRIT missiles, as well as the Roketsan MAM-L and MAM-C air-to-ground munitions. The Anka-B is in service with the Turkish Navy, General Directorate of Security and Gendarmerie General Command. The Anka-S in service with the Turkish Air Force and Gendarmerie General Command. The National Intelligence Agency (MIT) operates the Anka-I. As of mid-2020, a total of 23 Anka aircraft were in service with the Turkish military. The Tunisian Air Force has reportedly agreed to purchase multiple Anka-S aircraft.

Specifications

	Anka–B
Class	III
Status	A/A
Unveiled	2010 (Anka–A)
Operators	Tunisia, Turkey
Length	8.6m (28ft)
Width (wingspan)	17.5m (57ft)
Height	3.3m (10.7ft)
Maximum take-off weight	1,700kg (3,740lb)
Speed (cruise)	
Range (LOS)	250km (155 miles)
Endurance	30 hours
Flight ceiling	
Payload	350kg (770lb)
Armament	CIRIT, MAM-C, MAM-L, UMTAS

BEKDIL, B. E., 'Turkey's TAI sells six Anka-S drones to Tunisia', *Defence News*, 16 March 2020, https://www.defencenews.com/unmanned/2020/03/16/turkeys-tai-sells-six-anka-s-drones-to-tunisia/

DONALD, D., 'Turkish Aero unveils the Anka from Ankara', *Aviation International Online*, 20 July 2010, https://www.ainonline.com/aviation-news/defence/2010-07-20/turkish-aero-unveils-anka-ankara

MÖNCH PUBLISHING GROUP, 'Recent Developments in the Anka UAS Programme', *Monch Publishing Group*, 16 May 2018, https://www.monch.com/mpg/news/air/3356-anka-dev.html

SHEPHARD NEWS TEAM, 'Turkish SSM to purchase Anka UAV', *Shephard News*, 6 January 2012, https://www.shephardmedia.com/news/uv-online/ssm-pur-chase-anka-uav/

STEVENSON, B., 'Upgraded Anka carries out maiden flight', *Flight Global*, 2 February 2015, https://www.flightglobal.com/civil-uavs/upgraded-anka-carries-out-maiden-flight/115789.article

TURKISH AEROSPACE, 'Anka Successfully Completes Acceptance Tests', news release, 2013, https://www.tusas.com/en/news/anka-successfully-completes-acceptance-tests

TURNBULL, G., 'Turkey's classified SIGINT UAV breaks cover', *Shephard News*, 24 March 2018, https://www.shephardmedia.com/news/digital-battlespace/turkeys-classified-sigint-uav-breaks-cover/

This Anka exhibited at the Teknofest in September 2021 is armed with MAM–L missiles. (Stijn Mitzer and Joost Oliemans)

Vestel Defence Karayel

The Karayel is a Class III fixed-wing UCAV developed by Vestel Defence. It features high-mounted wings, a tractor propeller, and a conventional tailplane. It is available in both armed and unarmed configurations. In 2011, the Presidency of Defence Industries (SSM), Turkey's defence armaments directorate, awarded Vestel a contract to produce six Karayels. In 2014 and 2015, the Karayel conducted a series of flight tests ahead of delivery to the Turkish military. The Karayel conducted its first weapons trials in mid-2016 at Konya Air Base. Vestel Defence unveiled the Karayel-SU, a strike-capable extended-wing variant of the Karayel, at the 2017 Dubai Air Show. The payload capacity of the Karayel-SU is around 50kg (110lb) greater than the Karayel. At the IDEF 2019 defence exhibition, Vestel Defence displayed the Karayel-SU with the CIRIT, MAM-C, and MAM-L air-to-surface missiles. Limited numbers of Karayel UAVs are believed to be in service with the Turkish and Saudi militaries. In early 2020, the Houthi group in Yemen downed a Saudi-operated Karayel near the port of Al-Salif. In April 2020, the General Authority for Military Industries, Saudi Arabia's defence procurement agency, announced on Twitter that the Saudi military would acquire six Vestel Karayel-SU – known as the Haboob (بوباه) – UAVs in 2021 from the Saudi firm Intra Defense Technologies. It reportedly will acquire another 40 Karayel-SUs over the following five years.

Specifications

	Karayel–SU
Class	III
Status	A/A
Unveiled	2017
Operators	
Length	6.5m (21.3ft)
Width (wingspan)	13.0m (42.6ft)
Height	2.11m (6.9ft)
Maximum take-off weight	630kg (1,389lb)
Speed (cruise)	
Range (operational)	200km (124 miles)
Endurance	8-12 hours
Flight ceiling	
Payload (external)	120kg (264lb)
Armament	MAM-C, MAM-L

BINNIE, J., 'Karayel UAV lost over Yemen', *Janes*, 2 January 2020, https://www.janes.com/defence-news/news-detail/karayel-uav-lost-over-yemen

FORRESTER, C., 'Saudi Arabia announces UAV procurement', *Janes*, 28 April 2020, https://www.janes.com/defence-news/news-detail/saudi-arabia-announces-uav-procurement

MEVLUTOGLU, A. (@orko), 'Vestel Karayel tactical UAV scoring direct hit with ROKETSAN MAM-L', Twitter, 16 June 2016, https://twitter.com/orko_8/status/743393219518541825

AL-OHALI, A.A. (@Ahmad_A_AlOhali), https://twitter.com/Ahmad_A_AlOhali/status/1254789549634449410

OPALL-ROME, B., 'Turkey's newest armed drone makes debut at Dubai Airshow', *DefenseNews*, 15 November 2017, https://www.defencenews.com/digital-show-dailies/dubai-air-show/2017/11/15/turkeys-newest-armed-drone-makes-debut-at-dubai-airshow/

VESTEL, https://www.vestelsavunma.com/en/products/unmanned-aerial-vehicle-systems-p

A Karayel-SU equipped with MAM-C missiles on display at IDEX. (Georg Mader)

UKRAINE

Athlon ST-35

The ST-35 Silent Thunder is a hybrid loitering munition designed by Ahtlon Avia. It features a unique design comprising two air vehicles. The larger aircraft, which is equipped with the explosive payload, features an X-layout comprising four wings and four vertical stabilizers. The smaller aircraft is a hexacopter, a six-rotor vertical take-off and landing UAV. The hexacopter powers the primary air vehicle upward for launch to an altitude of 400 to 600m (1,312 to 1,969ft) before separating from the main air vehicle and acting as a signal relay for the operator. The primary air vehicle transitions to horizontal flight for the remainder of its journey. Athlon Avia unveiled the ST-35 in 2020.

Specifications

Class	Loitering munition
Status	B
Unveiled	2020
Operators	
Length	
Width (wingspan)	
Height	
Maximum take-off weight	9.5kg (21lb)
Speed (cruise)	140km/h (75.6kts)
Range	30km (19 miles)
Endurance	1 hour
Flight ceiling	1,200m (3,900ft)
Payload	3.5kg (8lb)
Armament	

ST-35 loitering munition. (GRIM)

'ST-35 Silent Thunder', Athlon Avia, accessed on 23 January 2021,
https://athlonavia.com/en/st-35-silent-thunder/

CDET RAM

The RAM is a fixed-wing loitering munition developed by CDET. It features high-mounted wings, a slender fuselage, and a catapult launch system. The RAM can be equipped with a range of explosive payloads such as a thermobaric, high-explosive anti-tank, or high-explosive fragmentation warhead. CDET and Spetstechnoexport, Ukraine's export directorate, unveiled the RAM in 2018 at the UMEX defence exhibition. It was displayed at the IDEX 2019 and UMEX 2020 shows.

Specifications

Class	Loitering munition
Status	B
Unveiled	2018
Operators	
Length	1.8m (5.8ft)
Width (wingspan)	2.3m (7.5ft)
Height	
Maximum take-off weight	8kg (18lb)
Speed (max)	150km/h (81kts)
Range	30km (19 miles)
Endurance	0.6 hours
Flight ceiling	
Payload	3kg (7lb)
Armament	

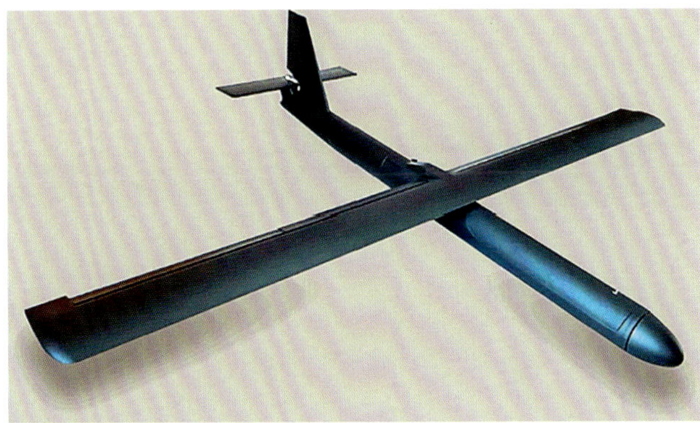

A drawing of the RAM. (CDET)

Unian Staff, 'Ukraine showcases kamikaze-style RAM UAV at UMEX 2018 exhibit in UAE', UNIAN, 27 February 2018,
https://www.unian.info/economics/10023710-ukraine-showcases-kamikaze-style-ram-uav-at-umex-2018-exhibit-in-uae.html

'Ram Uav', CDET, accessed on 22 February 2021,
https://ramuav.com/#partners

Novichkov, N., 'Ukraine unveils new RAM loitering munition', *Janes*, 9 March 2020,
https://www.janes.com/defence-news/news-detail/ukraine-unveils-new-ram-loitering-munition

Luch Sokil-300

The Sokil-300 is a conceptual Class III fixed-wing UCAV developed by Luch Design Bureau. It features high-mounted wings, a pusher propeller configuration, and conventional V-tail design. It is Ukraine's first domestic design for the medium-altitude long-endurance unmanned aircraft. Luch Design Bureau unveiled a full-scale mock-up of the Sokil-300 in a November 2020 event in Kyiv, where it was displayed fitted with two missile-launchers under each wing. According to the Luch Design Bureau, the Sokil-300 will be compatible with the 37kg (82lb) RK-2P anti-tank guided missile or the new 38kg (84lb) RK-10 missile. A subscale model of the Sokil-300 was displayed internationally for the first time at the IDEX 2021 defence exhibition. As of this writing, the Sokil-300 remains in development, and it does not appear as though it has conducted flight or weapons trials.

Specifications (if Rotax 914 equipped)

Class	III
Status	C/C
Unveiled	
Operators	2020
Length	8.6m (28ft)
Width (wingspan)	14m (46ft)
Height	
Maximum take-off weight	1,225kg (2,700lb)
Speed (max)	335km/h (180.9kts)
Range	
Endurance	26 hours
Flight ceiling	
Payload	300kg (660lb)
Armament	(RK-2P), (RK-10)

Sokin-300 with RK-2P missiles. (via Georg Mader)

Phillips, P., 'New Drones as Part of Ukraine's Modernisation Project', FINABEL, 13 November 2020, https://finabel.org/new-drones-as-part-of-ukraines-modernisation-project/?utm_source=rss&utm_medium=rss&utm_campaign=new-drones-as-part-of-ukraines-modernisation-project

'Ukraine's brand new combat UAV debuts at IDEX 2021 (Photo)', UNIAN, 21 February 2021, https://www.unian.info/economics/idex-2021-ukraine-showcases-brand-new-combat-uav-11329019.html

UADCOM SkyFist

The SkyFist is a prototype Class I fixed-wing UCAV developed by Ukranian Armed Drones Company (UADCOM), a startup founded by Ukrainian military veterans. It features a short, slender fuselage with high-mounted wings and a conventional T-shaped tailplane. It employs a single engine in the puller configuration and a catapult for launch. According to the manufacturer, the SkyFist can be outfitted with three specially designed munitions, as well as a medical kit. The munitions include the Fist1, a 75mm (2.95in) high-explosive unguided bomb; Fist2, a high-explosive anti-tank unguided bomb; and the Fist3, a high-explosive incendiary bomb. Work on the SkyFist began in 2016 and it was revealed to the public for the first time in 2021. Video footage released by the manufacturer in July 2021 shows the SkyFist conducting flight and weapons trials.

Specifications

Class	I
Status	B/B
Unveiled	2021
Operators	
Length	
Width (wingspan)	
Height	
Maximum take-off weight	
Speed (max)	
Range	45km (28 miles)
Endurance	
Flight ceiling	
Payload	2.0kg (4.4lb)
Armament	

SkyFist ready to launch. (UADCOM)

Malyasov, D., 'Ukrainian startup unveils new family of armed drones', Defence-Blog, 13 July 2021,
https://defence-blog.com/ukrainian-startup-unveils-new-family-of-armed-drones/

'SkyFist', UADCOM, accessed on 31 August 2021,
https://skyfist.biz/#about

United Arab Emirates

ADASI QX-1/2/3/4

The QX family is a set of four UAVs and loitering munitions developed by ADASI, a subsidiarity of the Emirati defence conglomerate Edge Group. The QX-1, QX-2, and QX-3 are rotary-wing UAVs with a standard quadcopter design, while the QX-4 is a hybrid UAV with the capacity for both vertical take-off and landing and horizontal flight. The QX-1/2/3 have an endurance of less than one hour, while the QX-4 can fly for up to 90 minutes. ADASI describes all of the QX series as 'loitering munitions' and although each can be operated as expendable, the QX-2, QX-3 and QX-4 appear to be designed to be recovered after deploying precision-guided munitions on the target. For the purposes of the analysis in this book, the QX-1 is designated as a loitering munition, while the QX-2/3/4 are Class I UCAVs. With a payload capacity of 1.5kg (3.3lb), the QX-2 is capable of being equipped with one munition, while the heavier QX-3 and QX-4 can carry between two and four munitions. The exact type of munition is unclear, though ADASI describes it as a precision-guided munition with an accuracy of one meter and developed by another Edge subsidiary. According to an ADASI spokesperson, the QX-2/3/4 aircraft are capable of sharing targeting information with the munition and of operating in communications-denied environments. ADASI unveiled the QX family at the IDEX 2021 defence exhibition.

Specifications

	QX–1	QX–2	QX–3	QX–4
Class	Loitering munition	I	I	I
Status	C	C/C	C/C	C/C
Unveiled	2021	2021	2021	2021
Operators				
Length				
Width (wingspan)				
Height				
MTOW	3kg (7lb)	8kg (18lb)	20kg (44lb)	20kg (44lb)
Speed (max)				
Range	10km (6 miles)	20km (12 miles)	40km (25 miles)	40km (25 miles)
Endurance	0.4 hours	0.5 hours	0.8 hours	1.5 hours
Flight ceiling	2,000m (6,560ft)	2,000m (6,560ft)	2,000m (6,650ft)	2,000m (6,560ft)
Payload	0.5 kg (1.1lb)	1.5 kg (3.3lb)	6.0kg (13lb)	5.0kg (11lb)
Armament				

Edge, 'Edge Unveils First UAE-Made Family of Smart Loitering Munitions at IDEX 2021', news release, 21 February 2021, https://adasi.ae/news

Helou, A., 'Edge Group unveils Kamikaze drones at IDEX', *Defense News*, 22 February 2021, https://www.defensenews.com/digital-show-dailies/idex/2021/02/22/edge-group-unveils-kamikaze-drones-at-idex/

Woessner, F., https://twitter.com/FeWoessner/status/1363758790387064833

▶ Three QX Class I copters. (Georg Mader)

ADASI RW-24

The RW-24 is a fixed-wing loitering munition developed by ADASI, a subsidiary of the Emirati conglomerate Edge. ADASI markets the RW-24 as a 'smart loitering munition' that utilizes Global Navigation Satellite System (GNSS) to deliver precision effects. The RW-24 was unveiled at the 2019 Dubai Air Show. In a 2020 interview with *Defense News*, Edge CEO Faisal Al Bannai said that the first deliveries of the RW-24 to the UAE Armed Forces were scheduled to begin in the fourth quarter of 2020. At the IDEX 2021 defence exhibition, ADASI unveiled extended range and extended warhead variants of the RW-24 that offer the option of increasing the payload capacity from 8 to 13kg (18 to 29lb).

Specifications

RW-24. (ADASI)

Class	Loitering munition
Status	B
Unveiled	2019
Operators	
Length	
Width (wingspan)	
Height	
Maximum take-off weight	45kg (99lb)
Speed (cruise)	130km/h (70.2kts)
Range	100km (62 miles)
Endurance	2 hours
Flight ceiling	
Payload	8.0kg (18lb)
Armament	

Edge, 'Edge Unveils First UAE-Made Family of Smart Loitering Munitions at IDEX 2021', news release, 21 February 2021, https://adasi.ae/news

Helou, A., 'Top exec of UAE's Edge talks Israeli cooperation, IDEX preparations', *Defense News*, 29 September 2020, https://www.defensenews.com/interviews/2020/09/29/top-exec-of-uaes-edge-talks-israeli-cooperation-idex-preparations/

Turk, N., 'The UAE, a major US weapons buyer, is building its own defence capabilities', *CNBC*, 17 November 2019, https://www.cnbc.com/2019/11/18/dubai-air-show-uae-must-build-its-own-military-capabilities-edge-ceo-says.html

ADASI SHADOW-25 and -50

The Shadow is family of high-speed, fixed-wing loitering munitions developed by ADASI, a subsidiary of Edge Group. The Shadow-25 has a payload capacity of 25kg (55lb), while the Shadow-50 has a 50kg (110lb) capacity.

Aldasi Shadow-50 (IDEX catalouge)

ADCOM Yabhon United U-40

The Yabhon United 40 is a Class III fixed-wing UCAV developed by Adcom Systems. It features a unique S-shaped fuselage, tandem wing design, and twin-engine configuration. Adcom Systems unveiled a full-scale mock-up of the United 40 at the 2011 Dubai Air Show alongside Adcom's 30kg (66lb) Namrod guided glide bomb. The finished United 40 design was displayed at the 2013 Dubai Air Show. At the 2015 IDEX defence exhibition, Adcom unveiled a maritime variant of the United 40 fitted with a single lightweight torpedo under the fuselage. At the time, ADCOM said that the maritime variant, also known as the Block 6, could dispense anti-submarine sonobuoys. In December 2018, Algeria revealed that it had taken delivery of several United 40s. Video footage of the Algerian United 40s showed the aircraft equipped with what appeared to be 120mm mortar rounds and Namrod munitions.

Specifications

Class	III
Status	A/A
Unveiled	2011
Operators	Algeria
Length	11m (36ft)
Width (wingspan)	20m (65.6ft)
Height	
Maximum take-off weight	
Speed (cruise)	
Range	
Endurance	
Flight ceiling	
Payload	
Armament	Namrod, mortars

DEMPSEY, J. (@JosephHDempsey), 'Interesting that #Algeria Yabhon United 40 UAV weapon options include 120mm mortar rounds (ID?) in addition to apparent #UAE Namrod ASMs', Twitter, 20 December 2018, https://twitter.com/JosephHDempsey/status/1075896479523463169

DONALD, D., 'Adcom Unveils Innovative UAV at Dubai Air Show', Aviation International Online, 13 November 2011, https://www.ainonline.com/aviation-news/defense/2011-11-13/adcom-unveils-innovative-uav-dubai-air-show

SHAW-SMITH, P., 'Adcom Still Ambitious To Grow UAV Business', Aviation International Online, 17 June 2013, https://www.ainonline.com/aviation-news/aerospace/2013-06-17/adcom-still-ambitious-grow-uav-business

The first flight of an armed ADCOM United U-40 Block-5 took place on 3 July 2013. (via Georg Mader)

United Kingdom

MBDA Spectre

The MBDA Spectre is Class I hybrid fixed-wing and vertical-take-off and landing UCAV. MBDA unveiled the Spectre at the 2018 DVD exhibition in the United Kingdom. The Spectre is powered by an electric motor and will have an operational range of 10km (6.2 miles) and endurance of 1 hour. It will have a 25kg (55lb) payload capacity with a hollow bay that could be used to carry weapons or munitions or supplies to forward-deployed soldiers. According to Aviation International Online, MBDA designed the Spectre to be compatible with the Enforcer, a lightweight guided missile for soft-skinned and lightly armoured targets. MBDA is also considering outfitting the Spectre with a single 15kg (33lb) MMP anti-tank guided missile. As of this writing, the Spectre remains in development.

Specifications

Class	I
Status	C/C
Unveiled	2018
Operators	
Length	
Width (wingspan)	
Height	
Maximum take-off weight	
Speed (cruise)	
Range (operational)	10km (6 miles)
Endurance	1 hour
Flight ceiling	
Payload	25kg (55lb)
Armament	(MMP anti-tank guided missile)

MBDA Spectre (MBDA)

STEVENSON, B., 'MBDA Unveils Company-level UAV Concept', *Aviation International Online*, 21 September 2018, https://www.ainonline.com/aviation-news/defense/2018-09-21/mbda-unveils-company-level-uav-concept

Thales UK Watchkeeper-X

The Watchkeeper X (WKX) is a concept for a strike-capable variant of the Watchkeeper WK450, a Class II surveillance and reconnaissance UAV developed by Thales UK. The Watchkeeper WK450 is itself a derivative of the Israel Aerospace Industries Hermes 450 that was designed to meet the British Army's requirements for a medium-range medium-altitude UAV. The Watchkeeper WK450 was unveiled at the Farnborough air show in 2004. In 2009, a British official announced that it was exploring the possibility of arming the WK450, possibly with the Thales Lightweight Multirole Missile. Thales unveiled the Watchkeeper X at the London DSEI 2015 defence exhibition, which it said could be armed and was designed to meet the specifications of potential customers like France and Poland. According to Thales, the WKX could be equipped with effectors such as the Freefall Lightweight Multirole Munition (FFLMM), a member of the Thales Martlet Lightweight Multirole Missile family. Since then, Thales has displayed the Watchkeeper X equipped with the FFLMM at several defence exhibitions.

As of early 2021, the Watchkeeper X was not operational, though the unarmed Watchkeeper WK450 remains in active service with the British Army. The Watchkeeper X is a reportedly contender for the Gryf programme, a Polish military acquisition programme aimed at procuring a tactical medium-range armed UAV.

Specifications

	Watchkeeper WK450
Class	II
Status	A/C
Unveiled	2004
Operators	
Length	6.5m (21.3ft)
Width (wingspan)	10.9m (36ft)
Height	
Maximum take-off weight	485kg (1,067lb)
Speed (cruise)	142km/h (76.7kts)
Range	150km (93 miles)
Endurance	14 hours
Flight ceiling	4,876m (16,000ft)
Payload	
Armament	FFLMM

Watchkeeper X. (via @dpmhaus1/Twitter)

FLIGHTGLOBAL STAFF, 'Thales and Elop provide dual payloads for Watchkeeper', *FlightGlobal*, 12 December 2005,
https://www.flightglobal.com/thales-and-elop-provide-dual-payloads-for-watchkeeper/64450.article

HOYLE, C., 'UK could arm its Watchkeeper UAVs', *FlightGlobal*, 3 November 2009,
https://www.flightglobal.com/uk-could-arm-its-watchkeeper-uavs/90023.article

SABAK, J., 'DSEI 2015: Watchkeeper X – Export Variant for France and Poland', *Defence24*, 17 September 2015,
https://www.defence24.com/dsei-2015-watchkeeper-x-export-variant-for-france-and-poland

SABAK, J., 'Poland Takes Step to Procure Armed Gryf UAVs', *Defence24*, 1 December 2020,
https://www.defence24.com/poland-takes-step-to-procure-armed-gryf-uavs

STEVENSON, B., 'Farnborough: Watchkeeper manufacture nears completion as Thales eyes new developments', *FlightGlobal*, 13 July 2016,
https://www.flightglobal.com/civil-uavs/farnborough-watchkeeper-manufacture-nears-completion-as-thales-eyes-new-developments/121175.article

'Watchkeeper', News and Events, Army.mod.uk, last modified on 28 August 2020, accessed on 3 March 2021,
https://www.army.mod.uk/news-and-events/news/2020/08/watchkeeper/

United States of America

Area-I Altius-600

The Altius-600 is a fixed-wing loitering munitions developed by Area-I, a subsidiary of Anduril Industries. It is part of the ALTIUS – Air-Launched, Tube-Integrated Unmanned System – family of UAVs, which includes the Altius-500 and Altius-900. The Altius-600 features a slender fuselage, high-mounted wings, and an inverted V-tail. It employs a single engine in the pusher configuration and can be air- or ground-launched. Work on the ALTIUS began in 2011, when the U.S. Air Force awarded Area-I a contract to develop lightweight, tube-launched drones capable of conducting a variety of missions such as ISR and communications relay. Although initially developed for non-kinetic missions, Area-I currently advertises the Altius-600 as capable of conducting strikes. The Altius-600 is one of the aircraft involved in the U.S. Army's Air Launched Effects (ALE) program, which seeks to develop drones and loitering munitions that can be launched from crewed rotary- and fixed-wing aircraft. In service of this program and others within the U.S. military, the Altius-600 has been tested on a variety of crewed and uncrewed aircraft, including the C-130 transport aircraft, UH-60 Black Hawk helicopter and MQ-1C Gray Eagle UCAV.

Specifications

Class	Loitering munition
Status	B
Unveiled	2017
Operators	USA
Length	1.0m (3.3ft)
Width (wingspan)	2.54m (8.3ft)
Height	
Maximum take-off weight	12.3kg (27lb)
Speed (max)	166km/h (89.6kts)
Range	440km (273 miles)
Endurance	4 hours
Flight ceiling	
Payload	2.7kg (6.0lb)
Armament	

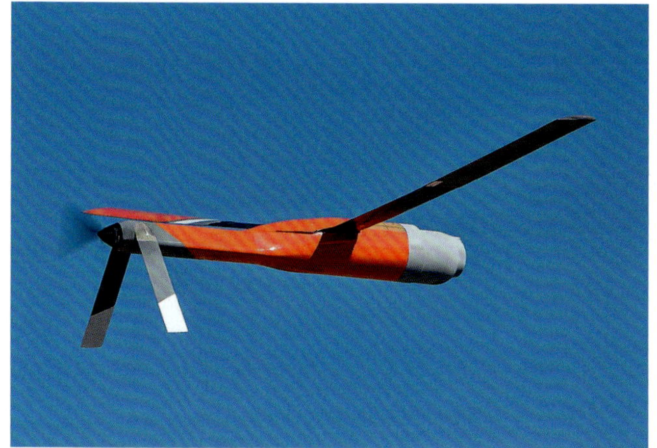

Area-I Altius-600. (CCDC/Jose Mejia-Betancourth)

AeroVironment Switchblade 300/600

The Switchblade is a family of fixed-wing loitering munitions developed by AeroVironment. The Switchblade features a tubular fuselage with low-mounted, foldable tandem wings and twin vertical stabilizers. It is designed to be tube-launched, though AeroVironment also offers a vehicle-mounted multipack launcher. The Switchblade 300, which was then known simply as the Switchblade, made its public debut at the 2011 AUVSI defence exhibition, though Bloomberg News reported in 2010 that the US Army had already conducted a limited deployment of the Switchblade in Afghanistan. The Army awarded AeroVironment a USD 4.9 million contract for the Switchblade in 2011 and the Air Force purchased it the following year. Since then, various US military services have acquired the Switchblade under the Lethal Miniature Aerial Munitions procurement project. In 2015, the US Marine Corps conducted a test in which it launched a Switchblade 300 from an MV-22 Osprey. In 2021, the UK announced that it would acquire the Switchblade 300, becoming the first known foreign operator of the system.

In October 2020, AeroVironment unveiled the Switchblade 600, a larger, heavier variant of the Switchblade that can fly farther and faster than its predecessor. The Switchblade 600 is equipped with an anti-armour warhead, which AeroVironment has said weighs approximately five times as much as that on the 300. As of late 2020, AeroVironment has conducted at least 60 test flights of the Switchblade 600. In April 2021, US Special Operations Command awarded AeroVironment a USD 26.1 million contract for an unspecified number of Switchblade 600s.

Specifications

	Switchblade 300	Switchblade 600
Class	Loitering munition	Loitering munition
Status	A	A
Unveiled	2011	2020
Operators	UK, USA	USA
Length		1.3m (4.3ft)
Width (wingspan)		
Height		
MTOW	2.5kg (6lb)	22.7kg (50lb)
Speed (cruise)	100km/h (54kts)	112km/h (60.5kts)
(maximum)	160km/h (86.4kts)	185km/h (100kts)
Range	10km (6 miles)	80km (50 miles)
Endurance	0.25 hours	0.6 hours
Flight ceiling		
Payload (warhead)	0.5kg (1.1lb)	2.5kg (5.5lb)
Armament		

Switchblade 300. (Tyler Fort)

Capaccio, T., 'US Deploys Kamikaze Drones to Attack Afghan Taliban Targets', *Bloomberg*, 19 October 2011,
https://www.bloomberg.com/news/articles/2011-10-19/u-s-flew-kamikaze-drones-to-attack-taliban/

Gettinger, D., 'Summary of Drone Spending in the FY 2019 Defence Budget Request', *Centre for the Study of the Drone at Bard College*, April 2018,
https://dronecenter.bard.edu/projects/defence-spending-on-drones/drones-in-the-fy-2019-defence-budget/

Lamothe, D., 'Marines launch a 'kamikaze' drone from an Osprey aircraft', *Washington Post*, 17 April 2015,
https://www.washingtonpost.com/news/checkpoint/wp/2015/04/17/marines-launched-a-kamikaze-drone-from-an-osprey-aircraft/

Reim, G., 'AeroVironment unveils anti-armour Switchblade 600 loitering munition', *FlightGlobal*, 1 October 2020,
https://www.flightglobal.com/military-uavs/aerovironment-unveils-anti-armour-switchblade-600-loitering-munition/140409.article

Reim, G., 'UK to buy Switchblade loitering munition', *FlightGlobal*, 17 March 2021,
https://www.flightglobal.com/military-uavs/uk-to-buy-switchblade-loitering-munition/142944.article

Reim, G., 'US Special Ops buys AeroVironment's anti-armour Switchblade 600 loitering munition', *FlightGlobal*, 27 April 2021, https://www.flightglobal.com/military-uavs/us-special-ops-buys-aerovironments-anti-armour-switchblade-600-loitering-munition/143483.article

Rizzo, J., 'The drone with a death wish', *CNN*, 19 October 2011, https://security.blogs.cnn.com/2011/10/19/the-drone-with-a-death-wish/

Shephard News Team, 'AeroVironment wins $4.9m US Army contract', *Shephard News*, 2 September 2011, https://www.shephardmedia.com/news/uv-online/aerovironment-wins-49m-us-army-contract/

Warwick, G., 'USAF Buys Switchblade Lethal Mini-UAVs', *Aviation Week*, 23 February 2012, https://aviationweek.com/defence-space/usaf-buys-switchblade-lethal-mini-uavs

GA-ASI MQ-1 Predator C/Avenger

The Avenger is a Class III fixed-wing UCAV developed by General Atomics Aeronautical Systems Incorporated (GA-ASI). Also known as the Predator C, the Avenger features a low-observable design with mid-mounted, swept-back wings. GA-ASI has marketed the Avenger as a multi-mission platform with greater penetrating strike capability in high-threat environments than other Predator-series UCAVs. Unlike the Reaper and others, the Avenger is equipped with top-mounted Pratt & Whitney PW545B turbofan engine, enabling it to reach a maximum speed of around 740 km/h (399kts), or 300 km/h (162kts) more than its predecessors. An Avenger prototype conducted its maiden flight in 2009. The Avenger Extended Range, an upgraded version with a longer wingspan and greater payload capacity, flew for the first time in 2016. A limited number of Avengers are believed to be in service with the US military, though the extent to which it has been deployed is unclear. In 2011, the US Air Force announced that it would deploy a demonstrator Avenger to Afghanistan for limited operational trials. In 2018, the Avenger served as the basis for GA-ASI's submission for the US Navy's MQ-25 Stingray tanker UAV, a competition that it lost to Boeing. In September 2020, GA-ASI unveiled the Defender, a concept UCAV that appears to be a based heavily on the Avenger and oriented towards air-to-air missions. In December 2020, GA-ASI announced that the Avenger had conducted an autonomous flight in support of the Air Force's Skyborg programme.

The Avenger has six external hardpoints and an internal weapons bay, yielding a combined payload capacity of nearly 3,000kg (6,614lb). According to General Atomics, it is reportedly compatible with the AGM-114 Hellfire guided missile and the GBU-12/49, GBU-31 GBU-32, GBU-38 JDAM, GBU-39, and GBU-16/48 guided bombs. However, the extent of the Avenger's weapons trials is unclear, and it is not obvious that each of these weapons have been tested with the Avenger as of this writing. In 2021, General Atomics announced that it had integrated the Lockheed Martin Legion Pod infrared search and track system onto an Avenger, indicating that GA-ASI could be preparing the Avenger for air-to-air missions. Under auspices of the ongoing MQ-Next programme, the Air Force has expressed a particular interest in a UCAV with an air-to-air capability as a potential replacement for the MQ-9 Reaper.

Predator-C Avenger. (GA)

Specifications

	Avenger ER
Class	III
Status	A/B
Unveiled	2009
Operators	
Length	13.0m (43ft)
Width (wingspan)	20.0m (66ft)
Height	
Maximum take-off weight	8,255kg (18,199lb)
Speed (maximum)	740km/h (399.6kts)
Range	
Endurance	20 hours
Flight ceiling	
Payload (total)	2,948kg (6,500lb)
Armament	(AGM-114, GBU-12/49, GBU-31 GBU-32, GBU-38 JDAM, GBU-39, GBU-16/-48)

CAREY, B., 'General Atomics Predator C Avenger ER Makes First Flight', *Aviation International Online*, 11 November 2016, https://www.ainonline.com/aviation-news/defence/2016-11-11/general-atomics-predator-c-avenger-er-makes-first-flight

DREW, J., 'General Atomics Claims 'Huge Advantage' Over Big Rivals For MQ-25', *Aviation Week*, 2 April 2018, https://aviationweek.com/defence-space/general-atomics-claims-huge-advantage-over-big-rivals-mq-25

GENERAL ATOMICS, 'GA-ASI Integrates Legion Pod Onto Avenger RPA', news release, 15 January 2021, https://www.aviationpros.com/aircraft/unmanned/press-release/21206124/general-atomics-aeronautical-systems-inc-gaasi-gaasi-integrates-legion-pod-onto-avenger-rpa

GENERAL ATOMICS AERONAUTICAL, 'Predator C Avenger', accessed on 12 March 2021, https://www.ga-asi.com/remotely-piloted-aircraft/predator-c-avenger

NEWDICK, T., 'Air Force's MQ-9 Reaper Drone Replacement Requirements Now Include Air-To-Air Combat Capability', *The Drive*, 9 March 2021, https://www.thedrive.com/the-war-zone/39677/mq-9-reaper-replacement-requirements-now-include-air-to-air-capability-in-contested-air-space

REIM, G., 'General Atomics shows off Defender UAV concept to protect refuelling tankers', *FlightGlobal*, 3 March 2020, https://www.flightglobal.com/military-uavs/general-atomics-shows-off-defender-uav-concept-to-protect-refuelling-tankers/137062.article

SHEPHARD NEWS TEAM, 'First flight for Predator C', *Shephard*, 20 April 2009, https://www.shephardmedia.com/news/uv-online/first-flight-for-predator-c/

STEVENSON, B., 'Avenger set for Afghanistan deployment', *Shephard News*, 13 December 2011, https://www.shephardmedia.com/news/uv-online/avenger-set-afghanistan-deployment/

TREVITHICK, J., 'General Atomics Avenger Drone Flew An Autonomous Air-To-Air Mission Using An AI Brain', *The Drive*, https://www.thedrive.com/the-war-zone/37973/general-atomics-avenger-drone-flew-an-autonomous-air-to-air-mission-using-an-ai-brain

GA–ASI MQ–1C Gray Eagle

The MQ-1C Gray Eagle is a Class III fixed-wing UCAV developed by General Atomics Aeronautical Systems Incorporated (GA-ASI). The Gray Eagle – previously known as the Warrior or Sky Warrior – features low-mounted wings, a bulging nose, and an inverted V-tail. It is part of General Atomics' Predator family of UCAVs, which includes the MQ-1 Predator, MQ-9 Reaper, and Predator C Avenger, among others. In late 2001, the US Army launched the Extended Range/Multipurpose (ER/MP) UAV programme to acquire a replacement for the IAI/Northrop Grumman R/MQ-5 Hunter. In 2003, the Army acquired a limited number of GA-ASI Improved GNAT UAVs, which it dubbed the Warrior-Alpha, as an interim solution while the ER/MP programme continued to evaluate solutions. In October 2004, GA-ASI conducted the maiden flight of a pre-production Warrior prototype, which would serve as the basis for GA-ASI's submission for the ER/MP programme. The following year, the Army awarded General Atomics a contract for an initial 17 Warrior aircraft and associated equipment for the ER/MP programme. In 2007, GA-ASI conducted the first flight tests of the production version of the Sky Warrior. In 2008, the US Army deployed the Sky Warrior to Iraq for the first time. In 2009, the Army conducted the Sky Warrior's initial weapons trials with inert HELLFIRE II air-to-ground missiles. In 2010, the Army settled on the name 'Gray Eagle' for the ER/MP system, discarding the Warrior and Sky Warrior monikers. In 2013, GA-ASI launched flights tests of the Improved Gray Eagle/Gray Eagle Extended Range (GE-ER), a heavier variant of the Gray Eagle that offered increased payload capacity and endurance. Today, the US Army operates around 200 MQ-1C Gray Eagles; it is not known to be flown by any other country.

The Gray Eagle's primary armament to date has been the AGM-114 Hellfire missiles, of which it can carry up to four. In May 2016, a US Army Gray Eagle conducted weapons trails with the AGM-179 Joint Air-to-Ground missile (JAGM), the US military's planned successor to the Hellfire. In 2019, US Army Special Operations Aviation Command conducted the Gray Eagle's first tests with the Dynetics GBU-69B Small Glide Munition, one of the longest-range bombs of its kind. Also in 2019, the Army announced that it was working on equipping the Gray Eagle with Air Launched Effects (ALE), the Army's umbrella term for drones such as loitering munitions or ISR UAVs that can be launched from aircraft instead of from the ground. In 2020, GA-ASI conducted a test in which it launched an Area-I Altius-600 drone from an MQ-1C Gray Eagle. According to Area-I, the Altius-600 is a small, fixed-wing UAV that can be equipped with kinetic, signals intelligence, or ISR payloads and has a range of over 400km (249 miles).

An MQ-1C from the 10th Aviation Regiment at Al-Asad Air Base for Operation Inherent Resolve. (US Army/SSG Isolda Reyes)

Specifications

	MQ-1C GE-ER
Class	III
Status	A/A
Unveiled	2004
Operators	USA
Length	9.0m (29.5ft)
Width (wingspan)	17.0m (55.8ft)
Height	
Maximum take-off weight	1,905kg (4,191lb)
Speed (maximum)	309km/h (167kts)
Range	4,630km (2,877 miles)
Endurance	42 hours
Flight ceiling	8,839m (29,000ft)
Payload (external)	227kg (500lb)
Armament	AGM-114, AGM-179 , Altius-600, GBU-69B

FLIGHTGLOBAL STAFF, 'GAASI flies first Sky Warrior UAV', *FlightGlobal*, 18 June 2007,
https://www.flightglobal.com/gaasi-flies-first-sky-warrior-uav/74291.article

GENERAL ATOMICS, 'US Army Procures IGNAT UAV System', news release, 27 May 2003,
https://www.ga-asi.com/u-s-army-procures-ignat-uav-system

GENERAL ATOMICS, 'Warrior™ HFE Demonstrator Unmanned Aircraft Takes Flight', news release, 25 October 2004,
https://www.ga-asi.com/warrior-hfe-demonstrator-unmanned-aircraft-takes-flight

GENERAL ATOMICS, 'General Atomics Aeronautical Systems and its Team Warrior Partners Awarded Major US Army Contract', news release, 10 August 2005, https://www.ga-asi.com/general-atomics-aeronautical-systems-and-its-team-warrior-partners-awarded-major-u-s-army-contract

GENERAL ATOMICS, 'First Sky Warrior UAS Deployed To Support Ground Forces In Iraq', news release, 12 June 2008,
https://www.ga-asi.com/first-sky-warrior-uas-deployed-to-support-ground-forces-in-iraq

GENERAL ATOMICS, 'Gray Eagle Extended Range (GE-ER)', Products, accessed on 12 March 2021,
https://www.ga-asi.com/remotely-piloted-aircraft/gray-eagle-extended-range

GOURLEY, S., 'AUVSI: It's Official: 'Gray Eagle'', *Shephard News*, 24 August 2010,
https://www.shephardmedia.com/news/uv-online/auvsi-its-official-grey-eagle/

HUGHES, R., 'US ARSOAC/USSOCOM trials GBU-69/B with Gray Eagle UAS', *Janes*, 7 October 2019,
https://www.janes.com/defence-news/news-detail/us-arsoacussocom-trials-gbu-69b-with-gray-eagle-uas

JUDSON, J., 'Joint Air-to-Ground Missile Fired From Drone, a First', *DefenseNews*, 2 June 2016,
https://www.defencenews.com/home/2016/06/02/joint-air-to-ground-missile-fired-from-drone-a-first/

REIM, G., 'General Atomics ready to add SparrowHawk attritable drone', *FlightGlobal*, 15 November 2019,
https://www.flightglobal.com/analysis/general-atomics-ready-to-add-sparrowhawk-attritable-drone/135249.article

REIM, G., 'General Atomics launches Altius-600 drone from MQ-1C Gray Eagle', *FlightGlobal*, 29 June 2020,
https://www.flightglobal.com/military-uavs/general-atomics-launches-altius-600-drone-from-mq-1c-gray-eagle/139049.article

US Army, 'New unmanned aerial system tests advanced missile', news release, 18 February 2010,
https://www.army.mil/article/34575/new_unmanned_aerial_system_tests_advanced_missile

GA-ASI MQ-9A Reaper/Predator B

The MQ-9A Reaper is a Class III fixed-wing UCAV developed by General Atomics Aeronautical Systems Incorporated (GA-ASI). The MQ-9A Reaper, which is also known as the Predator B, features low-mounted wings, a conventional V-tail, retractable landing gear, and push-propeller configuration. The MQ-9A is part of the Predator family of UCAVs, which includes the MQ-1A Predator, MQ-1C Gray Eagle, and MQ-9B SkyGuardian/SeaGuardian. GA-ASI began developing the Predator B as an Industrial Research and Development programme in 1999 with the goal of designing a larger version of the MQ-1A Predator that could carry out a wider variety of missions. In the early 2000s, General Atomics partnered with what is today the NASA Armstrong Flight Research Centre to work on developing the Predator B for high-altitude Earth science missions, a programme known as Altair. A pre-production Predator B-001 conducted its first flight in February 2001. With an enlarged payload capacity and longer wingspan, the Predator B could accomplish the same mission as four Predator As. The first Predator B prototypes were delivered to the US Air Force for an initial evaluation in February 2002. In 2004, the Air Force Program Executive Officer approved Milestone B, authorizing the acquisition of 10 pre-production prototypes and four aircraft for first low-rate initial production.

The US Air Force renamed the Predator B 'Reaper' in 2006, introducing the aircraft that would become the mainstay of US counterinsurgency and counterterror operations. Air Combat Command stood up the first two operational Reaper combat air patrols in September 2007. In the years since, US Reapers has been deployed to at least 12 countries to support US and allied operations in Afghanistan, Iraq, and Syria and across east and west Africa, the Arabian Peninsula, and eastern Europe. In addition to the United States, the MQ-9A is operated by the France, Italy, and the United Kingdom and in an unarmed capacity by Spain. By 2016, the Reaper had accumulated over 1 million flight hours. In 2018, the Air Force retired the MQ-1 Predator and transitioned to an UCAV fleet consisting entirely of Reapers. The following year, the US Air Force's Reaper exceeded 2 million flight hours, over 90 per cent of which were in support of combat operations.

The US Air Force's MQ-9A Reaper is typically equipped with the AGM-114R air-to-ground missile and the GBU-12 Paveway II laser-guided bomb. Since its introduction, the Air Force has integrated other types of munitions onto the Reaper, an effort that has accelerated in recent years. In May 2008, an Air Force Reaper launched a GBU-49, a variant of the GBU-12 that uses both laser and GPS guidance to reach its target. In May 2017, the Air Force conducted the Reaper's first test of the GBU-38 JDAM, another GPS-guided bomb. In 2018, an Air Force official revealed that a Reaper armed with 'heat-seeking air-to-air missile' – likely an AIM-9X Sidewinder – had downed an aerial target in a test the prior year. Then, in September 2020, the Air Force announced that the Reaper had conducted its first air-to-air kill, confirming that the Reaper is compatible with the AIM-9X Sidewinder Block 2. Also in September 2020, a Reaper conducted a test equipped with eight Hellfire missiles, double the usual capacity. In addition to efforts by the US Air Force, other international Reaper operators have added munitions to the Reaper's loadout. The UK's Reapers, for example, have been equipped with the MBDA Brimstone air-to-ground missile. Efforts to diversify the Reaper's loadout are likely to continue and GA-ASI and the Air Force endeavour to expand the Reaper's mission profile. In 2020, General Atomics announced that it had successfully tested a dispenser pod that is compatible with the Common Launch Tube, which allow the Reaper to equip with additional munitions such as the GBU-44/B Viper Strike glide bomb, the AGM-176 Griffin missile, or small loitering munitions.

An MQ-9 Reaper over the Nevada Test and Training Range on 15 July 2019. (USAF/A1C William Rosado)

Specifications

	MQ–9A Reaper
Class	III
Status	A
Unveiled	2001
Operators	France, Italy, Spain, UK, USA
Length	11.0m (36ft)
Width (wingspan)	20.0m (66ft)
Height	3.8m (12.5ft)
Maximum take-off weight	4,763kg (10,500lb)
Speed (maximum)	450km/h (243kts)
Range	1,815km (1,150 miles)
Endurance	27 hours
Flight ceiling	15,240m (50,000ft)
Payload (external)	1,361kg (3,000lb)
Armament	AGM-114, AIM-9X, Brimstone, GBU–12, GBU–38 JDAM, GBU–49

McGowan, L.,, US Air Force, 'MQ-9 Reaper UAV drops first GPS-guided weapon', news release, 16 May 2008, https://www.wpafb.af.mil/News/Article-Display/Article/400874/mq-9-reaper-uav-drops-first-gps-guided-weapon/

Pawlyk, O., 'MQ-9 Reaper Drone Flies with Double Hellfire Missiles in New Test', *Military.com*, 30 September 2020, https://www.military.com/daily-news/2020/09/30/mq-9-reaper-drone-flies-double-hellfire-missiles-new-test.html

Trevithick, J., 'New Pods Will Allow Reaper Drones to Hunt Submarines and Launch Small Weapons and Drones', *The Drive*, 21 January 2021, https://www.thedrive.com/the-war-zone/38843/new-pods-will-allow-reaper-drones-to-hunt-submarines-and-launch-small-weapons-and-drones

U.S. Air Force, 432nd Wing, 'MQ-9 Reapers add to arsenal with first GBU-38 drop', news release, 8 May 2017, https://www.af.mil/News/Article-Display/Article/1175849/mq-9-reapers-add-to-arsenal-with-first-gbu-38-drop/

U.S. Central Command, 'Statement from US Central Command on attacks against U.S. observation aircraft' 16 June 2019, https://www.centcom.mil/MEDIA/STATEMENTS/Statements-View/Article/1877252/statement-from-us-central-command-on-attacks-against-us-observation-aircraft/

GA-ASI MQ-9B Skyguardian/Seaguardian/Protector RG-Mk1

The MQ-9B SkyGuardian is a demonstrator Class III fixed-wing UCAV developed by General Atomics Aeronautical Systems Incorporated (GA-ASI). The MQ-9B is a modernised variant of the MQ-9A Reaper. Work began on the MQ-9B, then known as the known as the Certifiable Predator B (CPB), in 2012 as an internal GA-ASI research programme that aimed to produce an aircraft that met NATO airworthiness standards. A Predator B Extended Range, which would serve as the basis for the CPB, conducted its first flight in February 2016. The first Certifiable Predator B conducted its maiden flight in November 2016. Although the MQ-9B largely resembles its predecessor, the SkyGuardian is heavier and has a longer wingspan, giving it a greater endurance and payload capacity. In July 2018, an MQ-9B conducted the first ever trans-Atlantic flight for an UAV. Also in 2018, GA-ASI unveiled a maritime surveillance variant of the SkyGuardian known as the SeaGuardian. Although the SkyGuardian remains in development, several countries have plans to acquire the system.

The UK announced in its 2015 Strategic Defence and Security Review that it will replace its fleet of Royal Air Force MQ-9A Reapers with 20 'Protectors', which were later revealed to be the MQ-9B SkyGuardian. The UK's Protectors are scheduled to begin active service in the early 2020s. At least four other countries – Australia, Belgium, the Netherlands and Taiwan – have either purchased or announced their intention to acquire the SkyGuardian/SeaGuardian.

According to GA-ASI, the SkyGuardian has four hardpoints under each wing and one centreline hardpoint. Although the MQ-9B's final loadout remains unclear, it is expected to be compatible with the same munitions that have been fitted on the MQ-9A. However, the combination of the MQ-9B's greater payload capacity and certain customer- and mission-driven requirements is likely to lead to an expansion of the MQ-9B's magazine over time. In 2019, for example, MBDA announced that the UK's Protectors would be equipped with Brimstone air-to-surface missiles. In October 2020, GA-ASI successfully integrated a self-protection system on the MQ-9B SkyGuardian that is able to detect incoming surface-to-air threats and dispel countermeasures such as flares or chaff.

In November 2020, GA-ASI conducted a series of tests in which a SeaGuardian launched multiple AN/SSQ-53G sonobuoys, a sonar system designed for anti-submarine warfare. The SeaGuardian can carry up to four sonobuoy dispenser system pods, each of which can carry 10 A-size or 20 G-size sonobuoys.

An RAF MQ-9B with Brimstone missiles on display at RIAT 2018. (Jack Hawkins)

Specifications

	MQ–9B SkyGuardian
Class	III
Status	B/B
Unveiled	
Operators (on order)	(Australia, Belgium, Taiwan, UK)
Length	11.7m (38ft)
Width (wingspan)	24.0m (79ft)
Height	
Maximum take-off weight	5,670kg (12,500lb)
Speed (maximum)	390km/h (210.6kts)
Range	
Endurance	40 hours
Flight ceiling	12,192m (40,000ft)
Payload (total)	2,177kg (4,800lb)
Armament	AGM-114 Hellfire, Brimstone-3, Paveway IV

Carey, B., 'General Atomics Flies Extended-Range 'Long Wing' Reaper', *Aviation International Online*, 25 February 2016, https://www.ainonline.com/aviation-news/defence/2016-02-25/general-atomics-flies-extended-range-long-wing-reaper

General Atomics, 'GA-ASI's Type-Certifiable Predator B Takes Flight', news release, 28 November 2016, https://www.ga-asi.com/ga-asis-type-certifiable-predator-b-takes-flight

General Atomics, 'GA-ASI Successfully Completes Self-Protection System Demo on MQ-9', news release, 22 January 2021, https://www.ga-asi.com/ga-asi-successfully-completes-self-protection-system-demo-on-mq-9

Shephard News Team, 'General Atomics demonstrates self-contained ASW sonobuoy dispensing from UAS', *Shephard News*, 20 January 2021, https://www.shephardmedia.com/news/uv-online/general-atomics-demonstrates-self-contained-asw-so/

Stevenson, B., 'Protector' UAV fleet to replace RAF Reapers', *FlightGlobal*, 5 October 2015, https://www.flightglobal.com/civil-uavs/protector-uav-fleet-to-replace-raf-reapers/118405.article

Kratos Valkyrie

The XQ-58 Valkyrie is a prototype Class III fixed-wing UCAV developed by Kratos Defense and Security Solutions. The Valkyrie features a low-observability design with short, swept-back wings, a trapezoidal fuselage, and two vertical stabilizers. It incorporates a turbofan engine and an internal weapons bay and wing mounts each with a payload capacity of 272kg (600lb). Work on the Valkyrie began in 2016, when the U.S. Air Force Research Laboratory awarded Kratos a contract to develop a stealthy, high-speed drone that could one day accompany crewed fighter jets into combat, a program known as the Low Cost Attritable Strike Demonstrator. Crucially, the Air Force sought to produce an aircraft affordable enough to be produced in large quantities, one that could act as a supplement to its fleets of exquisite – and expensive – crewed fighters. It also sought a system that could be launched and recovered without the use of an airfield – the XQ-58 is launched using rocket boosters and recovered by parachute. The Valkyrie conducted its first test flight in 2019. In 2021, the Air Force conducted a successful demonstration of the Valkyrie's internal weapons bay, launching an Area-I Altius-600 UAV.

Specifications

Class	III
Status	B/B
Unveiled	2019
Operators	
Length	9.2m (30ft)
Width (wingspan)	8.2m (27ft)
Height	
Maximum take-off weight	2,721kg (6,000lb)
Speed (max)	1,050km/h (567kts)
Range	3,000km (1,860 miles)
Endurance	
Flight ceiling	
Payload	544kg (1,200lb)
Armament	Altius-600

An XQ-58A Valkyrie at US Army Yuma Proving Ground, Arizona, on 9 December 2020. (USAF/SSG Joshua King)

Northrop Model 437

The Model 437 is a conceptual Class III fixed-wing UCAV developed by Northrop Grumman. An artist's rendering of the Model 437 revealed in September 2021 shows an aircraft with low-mounted, swept-back wings, a V-tail vertical stabilizer, and a turbofan engine. The Model 437 would be equipped with landing gear and require a 914m (3,000ft) long runway for takeoff and landing. The Model 437 design is based on that of the Model 401 Sierra, a crewed demonstrator aircraft developed by Northrop subsidiary Scaled Composites. According to Northrop, the Model 437 is intended to serve as a loyal wingman platform to crewed fighter aircraft by probing enemy airspace or by carrying additional munitions, possibly AIM-120 Advanced Medium-Range Air-to-Air Missiles. Northrop is reportedly considering offering the Model 437 for the Skyborg and Mosquito programs, two loyal wing research and development programs at the U.S. Air Force and U.K. Royal Air Force, respectively.

Specifications

Class	III
Status	C/C
Unveiled	2021
Operators	
Length	
Width (wingspan)	
Height	
Maximum take-off weight	
Speed (cruise)	652km/h (352kts)
Range	5,560km (3,450 miles)
Endurance	
Flight ceiling	
Payload	453kg (1,000lb)
Armament	

Model 437 concept. (Northrop Grumman)

Northrop MQ-8 Fire Scout

The MQ-8 Fire Scout is a Class III rotary-wing UAV developed by Northrop Grumman. The MQ-8A and MQ-8B versions of the Fire Scout are based on the crewed Schweizer Aircraft Model 330 helicopter, while the MQ-8C version is modeled on the larger Bell 407 helicopter. The Fire Scout made its maiden flight in 1999. Unarmed versions of the Fire Scout are in service with the U.S. Navy. In 2013, the Navy conducted land-based live fire tests with the MQ-8B and the Advanced Precision Kill Weapon System, a 70mm (2.75in) guided rocket munition. The Navy never deployed armed MQ-8Bs, however, and efforts to integrate the weapon on the MQ-8C appear to have been delayed. In 2021, the Navy announced that it was seeking a replacement for the MQ-8B/C. In seeking to expand the Fire Scout's potential mission set Northrop Grumman has conducted tests involving miniature sonobuoys launched from the Fire Scout, suggesting that the drone could be used to conduct anti-submarine warfare operations.

Specifications

	MQ-8C
Class	III
Status	A/B
Unveiled	1999 (MQ-8A)
Operators	
Length	11.7m (38.7ft)
Width (wingspan)	11.2m (36.7ft)
Height	3.3m (10.8ft)
Maximum take-off weight	2,721kg (5,986lb)
Speed (max)	250km/h (135kts)
Range	277km (173 miles)
Endurance	15 hours
Flight ceiling	4,876m (16,000ft)
Payload	318kg (700lb)
Armament	APKWS

Northrop MQ-8C Fire Scout. (USN/ENS Jalen Robinson)

Insinna, V., "Weaponization of unmanned Fire Scout helicopter 'on hiatus' until 2023', *Defense News*, 9 April 2018, https://www.defensenews.com/digital-show-dailies/navy-league/2018/04/09/weaponization-of-unmanned-fire-scout-helicopter-on-hiatus-until-2023/

Lagrone, S., 'Northrop Grumman Pitching Fire Scout Helicopter Drone for ASW Missions', *USNI News*, 16 February 2021, https://news.usni.org/2021/02/16/northrop-grumman-pitching-fire-scout-helicopter-drone-for-asw-missions

Stevenson, B., 'AUVSI 2013: MQ-8 tidies up APKWS integration', *Shephard News*, 15 August 2013, https://www.shephardmedia.com/news/uv-online/mq-8-tidies-apkws-integration/

Raytheon Coyote

The Coyote is a family of fixed-wing loitering munitions developed by Raytheon. The Coyote Block 1 features a cylindrical fuselage with tandem, foldable wings and twin vertical stabilizers. It employs a single brushless electric engine in the pusher configuration and a canister launch system. It was developed by Advanced Ceramics Research – then a subsidiary of BAE Systems that was later acquired by Raytheon – for the U.S. Navy and unveiled in 2006. Although it was designed as an expendable drone that could be fitted with an explosive payload, the Coyote has since served in a variety of capacities, including as a vehicle for hurricane research. In 2016, the U.S. Navy simultaneously launched more than two dozen Coyote Block 1s to test a framework for controlling swarms of drones. The Coyote Block 2 and Block 3 are high-speed variants of the Coyote that Raytheon developed for the U.S. Army's Howler counter-drone system. Both the Block 2 and 3 employ a single jet engine and were unveiled in 2019 and 2020, respectively. In 2020, the U.S. government approved the Coyote Block 2 for export, though it is unclear as of this writing if Raytheon has secured contracts to do so. In July 2021, the U.S. Army announced that Coyote Block 3s had successfully intercepted a swarm of 10 drones using 'air-to-air kinetic defeats' in a milestone for its counter-drone efforts. In August 2021, Raytheon announced that it had flight tested a variant of the Coyote that it has designed for the Army's Air Launched Effects (ALE) program, which seeks to develop drones and loitering munitions that can be launched from crewed rotary- and fixed-wing aircraft.

Specifications

	Coyote Block 1
Class	Loitering munition
Status	A
Unveiled	2006
Operators	USA
Length	0.9m (3ft)
Width (wingspan)	1.5m (4.9ft)
Height	
Maximum take-off weight	5.9kg (13lb)
Speed (max)	102km/h (55kts)
Range	200km (124 miles)
Endurance	1.5h
Flight ceiling	7,610m (25,000ft)
Payload	2.25kg (5lb)
Armament	

Coyote Block 1. (NOAA)

Martinic, G., 'Swarming, Expendable, Unmanned Aerial Vehicles as a Warfighting Capability', *Canadian Military Journal 93*, No. 2 (2020), available at
http://www.journal.forces.gc.ca/vol20/no4/PDF/CMJ204Ep43.pdf

Parsch, A., 'Advanced Ceramics Research Coyote', Directory of U.S. Military Rockets and Missiles, last updated in 2006,
http://www.designation-systems.net/dusrm/app4/coyote.html

Reim, G., 'Raytheon flight tests Coyote air-launched effect', *FlightGlobal*, 3 September 2021,
https://www.flightglobal.com/military-uavs/raytheon-flight-tests-coyote-air-launched-effect/145340.article

Tingley, B., 'Jet-Powered Coyote Drone Defeats Swarm In Army Tests', *The Drive*, 26 July 2021,
https://www.thedrive.com/the-war-zone/41689/latest-coyote-drone-variant-defeats-drone-swarm-in-new-army-tests

Textron RQ-7 Shadow/NightWarden

The RQ-7 Shadow is a Class II fixed-wing UAV developed by AAI Corporation, a subsidiary of Textron Systems. The Shadow is the successor to the Israeli-made IAI RQ-2 Pioneer, which made its combat debut in the 1980s. Since then, AAI has produced a series of improved versions of the Shadow. The basis for the Shadow that is currently in active service is the RQ-7B Shadow 200, which was unveiled in the late 1990s. As of this writing, all Shadows that are currently in active military service are unarmed and used for intelligence-gathering. However, in 2016, Textron Systems and Thales conducted a successful weapons trial involving a Shadow and a Fury lightweight precision guided munition, a derivative of the Thales Lightweight Multirole Missile. Unlike the Shadow, the NightWarden, a demonstrator Class II fixed-wing UCAV developed by AAI/Textron Systems, has been marketed explicitly as a strike-capable system. Designed as the successor to the Shadow, Textron unveiled the NightWarden at the 2017 Paris Air Show. The NightWarden has straight mid-mounted wings, twin booms, and an inverted V-tail. Textron displayed the NightWarden at the 2017 Association of US Army (AUSA) defence exhibition fitted with Textron Fury glide bombs. The NightWarden is not known to be in active military service with any country.

Specifications

	Shadow V2 Block III	NightWarden
Class	II	II
Status	A/B	B/B
Unveiled	1999	2017
Operators		
Length		3.9m (13ft)
Width (wingspan)	6.2m (20ft)	7.0m (23ft)
Height		
MTOW	243kg (535lb)	340kg (750lb)
Speed (maximum)		157km/h (84.8kts)
Range (LOS)	125km (78 miles)	125km (78 miles)
(SATCOM)		1,000km (620 miles)
Endurance	8 hours	15 hours
Flight ceiling	5,500m (18,000ft)	5,500m (18,000ft)
Payload	43kg (95lb)	
Armament	Fury	Fury

RQ-7B ready to launch. (Joe Copalman)

PHILLIPS-LEVINE, T., PHILLIPS-LEVINE, D. and MILLS, W., 'Unmanned, Lethal, and Organic: The Future of Air Support for Ground Combat Forces', *Modern War Institute*, 7 January 2020, https://mwi.usma.edu/unmanned-lethal-organic-future-air-support-ground-combat-forces/

SPRENGER, S., 'Textron angles to land first customer of its new 'Nightwarden' tactical drone', *DefenseNews*, 19 June 2017, https://www.defencenews.com/digital-show-dailies/paris-air-show/2017/06/19/textron-angles-to-land-first-customer-of-its-new-nightwarden-tactical-drone/

TEXTRON SYSTEMS, 'Textron Systems and Thales Successfully Test Fury Weapon Guidance Package', news release, 11 July 2016, https://www.textronsystems.com/our-company/news-events/articles/press-release/textron-systems-and-thales-successfully-test-fury

TEXTRON SYSTEMS, 'Shadow® Tactical Unmanned Aircraft Systems' https://www.textronsystems.com/products/shadow-tactical-unmanned-aircraft-systems

TREVITHICK, J., 'Highlights From the Showroom Floor at the Army's Biggest Arms Expo and Conference', *The Drive*, 12 October 2017, https://www.thedrive.com/the-war-zone/15030/highlights-from-the-showroom-floor-at-the-armys-biggest-arms-expo-and-conference

Afghanistan and Pakistan (2001–today)

Data
Aircraft: CH-3B, MQ-1 Predator, MQ-9 Reaper
Munitions: AGM-114 Hellfire, AR-1, GBU-12, AR-1

Pakistan

In 2015, Pakistan announced that one of its Burraq UCAVs had conducted a strike in North Waziristan province, killing three militants, according to a Pakistani military spokesperson.[15] It was the first reported strike by Pakistan's Burraq drones, a licensed variant of the Chinese Caihong-3B, and using the Barq missile, a variant of the Chinse AR-1 anti-tank guided missile. Since then, however, there have been no reported incidents of drone strikes by Pakistan, although the Burraq remains in active operations and Pakistan has repeatedly displayed it fitted with Barq missiles. In 2021, Pakistan reportedly took delivery of five CASC CH-4s, a significantly more capable UCAV than the CH-3, possibly signifying a more expansive role for UCAVs in future operations, though the status of these UAVs is not clear.[16]

United Kingdom

The UK deployed MQ-9A Reapers to Afghanistan in October 2007 to support Operation Herrick, the UK's contribution to the US-led NATO campaign against the Taliban in Afghanistan. Operation Herrick marked the UK's first operational deployment of the Reaper and of armed UAVs. Between 2007 and 2014, when the Reapers were withdrawn from Afghanistan for operations against the self-styled Islamic State in Iraq and Syria, the Royal Air Force's Reapers flew more than 5,200 missions – around 5,300 sorties – and accumulated over 71,000 flight hours, according to statistics published by the UK Ministry of Defence in 2015.[17] The Reaper was one of three fixed-wing RAF combat aircraft deployed to Afghanistan during this period; they were joined at various times by the Harrier and the Tornado, two crewed fighter and attack aircraft. Of the three, the RAF's Reapers accumulated twice as many flight hours on average per year than the two crewed aircraft. Over the course of their deployment, seven per cent of all Reaper missions in Afghanistan involved a weapons release. Precision guided munitions delivered by Reapers – Hellfire missiles and GBU-12 bombs – constituted 51 per cent of all PGMs expended by RAF combat aircraft between 2005 and 2014.

United States

In October 2001, one month after the September 11 attacks on the World Trade Centre towers in New York, the US conducted the first airstrike by a drone.[18] A US MQ-1 Predator

An armed Burraq (CH–3A) during a Pakistani parade. (via Georg Mader)

armed with Hellfire missiles had tracked a convoy carrying Mullah Mohammad Omar, the leader of the Taliban, to a compound in near Kandahar, Afghanistan. Though the strike missed Omar, the event marked the first operational deployment of armed drones. Over the course of the more than two decades that followed, drones would be a feature of US operations in Afghanistan against al-Qaeda and the Taliban, as well as other organisations such as local affiliates of the self-styled Islamic State. In September 2007, the MQ-9 Reaper made its operational debut in Afghanistan, followed shortly thereafter by the first US UCAV strike involving the GBU-12 guided bomb.[19]

As with any military operation and particularly one that has continued for two decades, statistics on UCAV operations in Afghanistan are not comprehensive. Based on the limited data that is publicly available, strikes by Coalition MQ-1 Predators and MQ-9 Reapers grew in the period following the initial Reaper deployment, rising from 257 weapon releases in 2009 to 506 in 2012, though strikes by RAF Reapers probably account for between one-quarter and one-third of these.[20] In the years that followed, some analysts have observed that the total number of airstrikes in Afghanistan has increased while sorties by crewed combat aircraft remained flat, an inconsistency that could be attributed to a rise in drone strikes, though UCAV-specific data for this period is not publicly available.[21] Other indicators suggest that UCAV operations in Afghanistan have intensified even as the US has prepared to withdraw from the country. In January 2018, the Air Force announced that Afghanistan's Kandahar Airfield was host to the largest MQ-9 Reaper deployment at a single air base in the world.[22] Meanwhile, in September 2018, the US Marine Corps joined the Air Force in supplying contractor-operated intelligence-gathering MQ-9 Reapers in Afghanistan.[23] The Marine Corps initiated plans in 2021 to acquire a pair of Reapers from US contractors, a necessary step to conduct UCAV strikes.[24]

One of the early challenges the US and its partners faced in securing Afghanistan was the fact that many – if not all – of the Taliban's leaders had fled to neighbouring Pakistan, specifically its remote northwest territories. Beginning in 2004, the US began using drones to locate and target members of the Taliban leadership in an effort that would, together with similar campaigns in Somali and Yemen, become known as the targeted killing campaign. The first known US drone strike killed Nek Muhammad, a Pakistani leader of the Taliban from the South Waziristan region.[25] Over the next few years, the US would conduct a limited number of drone strikes in Pakistan. In the *New Yorker*,

Steve Coll writes that the US Central Intelligence Agency, which was leading the US effort at the time, was reluctant to rely too heavily on the tactic out of a concern for the political stability of Pakistan's government.[26] The pace of strikes increased in 2008 and 2009 after President Bush approved a plan proposed by Michael Hayden, then the director of the CIA, to broaden the scope of the campaign and loosen targeting guidelines, a reaction to the increasing flow of Taliban fighters from Pakistan to Afghanistan, according to Coll. The CIA would begin conducting drone strikes without asking for approval from Pakistan's government and targeting lower-rank militants who appeared to be engaging in or supporting Taliban missions. The CIA's targeting of individuals whose identify was unknown would become known as 'signature strikes' and lead to criticism that the CIA was unjustly targeting individuals with insufficient information.

But perhaps the single event that contributed most to the pace of strikes occurred in December 2009, when a Jordanian-born Al Qaeda operative attacked a CIA installation in Khost, Afghanistan, killing several Agency personnel. After the attack, according to one unnamed Obama administration official who spoke with *Washington Post* journalist Joby Warrick, 'political sensitivities were no longer a reason not to do something.'[27] In May 2010, a US drone strike killed Mustafa Abu Al Yazid, the third-ranking Al Qaeda leader and commander of the group's operations in Afghanistan, though two women and a child were also killed in the attack.[28][29] According to the New America Foundation, a Washington D.C.-based think tank, the US conducted 122 strikes in Pakistan in 2010, the most of any year.[30] In mid-2011, concerned that drone strikes were contributing to deteriorating relations with Islamabad and facing increasing criticism from US military and diplomatic officials, the Obama administration temporary halted the strikes, and by the end of the year the CIA agreed to limit the aggressive campaign.[31] The last known US strike in Afghanistan occurred in 2018.[32] Over the course of the campaign, drone strikes in Pakistan are believed to have resulted in the deaths of a number of Al Qaeda and Taliban leaders, including Baitullah Mehsud, leader of the Pakistani Taliban, in 2009 and Atiyah Abd Al Rahman, Al Qaeda's deputy commander, in 2011.[33] Numerous investigations by non-governmental organisations have documented cases of civilian harm. Estimates for the number of non-combatant deaths vary by organisation, though most agree that a minimum of 300 to 400 civilians have been killed in Pakistan by the strikes.

Iraq (2014–today)

Data

Aircraft: Baykratar-TB2, CH-4, MQ-1 Predator, MQ-1C Gray Eagle, MQ-9 Reaper

Munitions: AGM-114 Hellfire, GBU-12, GBU-38, MAM-C, MAM-L

Iran/Iran-backed militias

Iran's military experience with drones in Iraq date back to the Iran-Iraq War in the 1980s, the origin of Iran's domestic drone manufacturing ambitions. In recent years, Iran has deployed drones over Iraq intermittently and largely against Kurdish militant organisations. In 2018, for instance, a drone operated by the Iranian Revolutionary Guard Corps (IRGC) monitored and assessed an Iranian ballistic missile strike targeting Iranian Kurdish militants in northern Iraq.[34] The following year, in July 2019, a report in an Iranian media outlet and in Reuters claimed that the IRGC had launched drone strikes against a Kurdish headquarters and training camp in Iraq's Kurdistan region.[35] Given the limited operational range of Iranian UAVs and the lack of Iranian airfields along its border with Iraq, Iranian military UAV operations in Iraq have been limited compared to those in, for example, the south-eastern part of the country and along its southern coast.

Unlike Iranian military drones, Iranian-backed militias have greater flexibility to strike from within Iraq. Indeed, the US has accused Iranian-backed militias of conducting drone attacks on targets in Iraq and in neighbouring Saudi Arabia. In June 2019, the *Wall Street Journal* reported that US military intelligence blamed Iranian-backed militias in Iraq for a drone attack the previous month on a Saudi oil pipeline, though the Houthi group in Yemen claimed credit for the attack.[36] Similarly, US officials suspect that the drones and missiles that struck the Abqaiq and Khurais oil processing facilities in Saudi Arabia in September 2019 were launched in Iraq or Iran, or from both countries.[37] In April 2021, Kurdish officials said that unidentified perpetrators had used an explosive-laden drone was used in an attack aimed at US troops based at the airport in Erbil, an attack claimed by an organisation with ties to Iranian-backed militias.[38] According to a US military intelligence assessment quoted in an investigator general's report, the attack involved fixed-wing drones that were 'shipped dissembled from Iran and then assembled and launched from Iraq.'[39] These events suggest that drones offer militia groups tactical advantages in terms of range and accuracy over other forms of attacks such as rockets and mortars. They also reflect the difficulty of identifying the perpetrators of the attack and, without additional photographic and forensic evidence, the source of the drones used in the attack.

Iraq

In 2015, Iraq's defence ministry published a video of a Chinese-made CASC CH-4B conducting a combat mission in western Iraq against the self-styled Islamic State.[40] Between 2015 and 2016 Iraq reportedly took delivery of up to 20 strike-capable CASC CH-4Bs. In the years that followed, however, Iraq's CH-4s appear to have had a limited impact on its campaign against IS. In a report in 2019, US government inspectors attributed the inactivity to Iraq's difficulty maintaining the Chinese-made aircraft, observing that only one CH-4 appeared to be active at the time.[41] Another report by US government inspectors in spring 2021 found that of Iraq's fleet of 20 CH-4s, 8 had crashed and 12 were inactive because Iraq was awaiting maintenance parts for the aircraft from China.[42] The last Iraqi CH-4 mission as of this writing occurred in September 2019, according to the US report.

Turkey

In mid-2019, Turkey launched Operation Claw (Pençe Harekatı), a cross-border campaign targeting members of the Kurdistan Worker's Party (PKK). Successive cross-border operations included Claw-2, Claw-Eagle and -Tiger in 2020, and Claw-Lightning and -Thunderbolt in 2021, the last of which remains ongoing as of this writing. Turkish UAVs and UCAVs are believed to have played a role in each of these operations and, although precise figures are not available, witness and observer accounts suggest that Turkish drones are familiar presence in northern Iraq. Speaking to Agence-France Presse in 2020, one local Kurdish politician remarked that '*Not a day goes by without us seeing a drone.*'[43] In an April 2021 account in the *Guardian*, Kurdish residents of northern Iraq recounted how the ubiquity of Turkish warplanes and drones had affected their lives.[44] Turkish drone strikes in Iraq have upset relations between the two countries, as well as between Washington and Ankara. In July 2020, Iraqi President Barham Salih condemned a reported Turkish drone strike that seriously injured multiple civilians in northern Iraq.[45] The following month, Iraqi officials criticised a 'blatant Turkish drone attack' in the autonomous Kurdish region that killed Iraqi border guards.[46] Speaking to *France24* in 2020, an unnamed US source said that the 'frequency and intensity' of Turkish drone strikes had sown 'mistrust and irritation.'[47]

United Kingdom

The UK launched Operation Shader, the name for its contribution to the campaign against the self-styled Islamic State in Iraq and Syria (ISIS), in September 2014. The Royal Air Force has, at various times, deployed up to 10 MQ-9 Reapers UCAVs, as well as crewed Tornado, Typhoon, and F-35 combat aircraft to conduct air operations in support of Operation Shader. As of this writing, RAF Reapers comprise around half of the combat and ISR aircraft deployed to Shader. The first Reaper operation in support of Shader occurred on 22 October and the first strike on 9 November 2014.[48] Since then, RAF MQ-9 Reapers have conducted over 2,100 missions in Iraq, approximately 45 per cent of all RAF combat and ISR sorties in the country, according to statistics released by the Ministry of Defence to DroneWarsUK, a London-based non-governmental organisation.[49] The RAF's Reapers have conducted 350 strikes in Iraq, approximately 24 per cent of the total, and released 678 weapons, either AGM-114 Hellfire missiles or GBU-12 guided bombs, approximately 21 per cent of the total number of weapons released by the RAF in Iraq.[50] Strikes by RAF Reapers in Iraq have declined from a peak in June 2016, when the Reaper conducted 25 strike sorties, releasing 49 AGM-114 Hellfires and 11 GBU-12 bombs. By comparison, of the 408 missions Reaper missions in Iraq over the course of 2020, only nine involved a weapons release. This corresponds to the overall progress of the counter-IS campaign in Iraq, which reached its apex in 2016 and 2017 with the Battle of Mosul. Total RAF Reaper flying hours in Iraq and Syria in support of Shader have likewise declined from nearly 13,000 in 2015 to around 7,200 in 2020, though they were higher in 2020 than at any other point in the previous four years. In 2020, RAF Reapers conducted approximately 55 per cent of all RAF air missions in Iraq and 49 per cent of all RAF missions in both Iraq and Syria.

United States

(This section reflects US counter-ISIS operations in both Iraq and Syria.) In August 2014, the US launched Operation Inherent Resolve (OIR), the official name for US combat operations against the self-styled Islamic State in Iraq and Syria (ISIS). An armed US Air Force UCAV participated in the opening salvo of the war, striking an ISIS target in northern Iraq with a Hellfire missile. At the outset of operations against ISIS, the US Air Force deployed armed MQ-1 Predators and MQ-9 Reapers, before transitioning to an MQ-9-only fleet in 2018. The Air Force's MQ-9 Block 5, the latest variant of the Reaper, made its combat debut in operations against ISIS in 2017, striking targets with a combination of Hellfire missiles and GBU-38 bombs, marking what was likely the first time that a Reaper employed a GBU-38 in combat.[51] The US Army deployed MQ-1C Gray Eagles, which are also capable of carrying Hellfire missiles, to Iraq at some point prior to efforts to retake the city of Mosul. It is difficult to know with certainty the extent to which Air Force and Army UCAVs participated in strike operations in Iraq over the length of the campaign, particularly since most data released by the US military is not specific

An Iraqi CH-4 ground control station. (Iraqi Ministry of Defence)

The armed Iraqi Air Force CH-4 'YI-801'. (Iraqi Ministry of Defence)

to either Iraq or Syria, nor to aircraft type. However, an analysis of the campaign by the RAND Corporation, a US research institute, found that US Predators and Reapers were 'the platforms in greatest demand in [Operation Inherent Resolve]'.[52] Given the limited numbers of US personnel on the ground, US Joint Tactical Air Controllers relied largely on full-motion video feeds from drones to monitor partner forces and locate targets for strikes, particularly at the outset of the campaign. As a result, US UAVs were 'overtasked relative to supply with too many missions,' contributing to shortfalls in capacity, according to RAND. US Air Force Predators and Reapers operated from bases in various locations, including Incirlik Air Base in Turkey, Muwaffaq Salti Air Base and H4 Air Base in Jordan, and Ali Al Salem Air Base in Kuwait.[53] Army MQ-1C Gray Eagles were based at Al Asad Air Base in Iraq.

The limited official data available suggests that Air Force armed drone operations in Iraq and Syria in support of OIR probably increased as the campaign intensified. In the first year of the war (August 2014 to August 2015), Air Force Predators and Reapers conducted 4,300 combat and ISR sorties and expended 1,000 munitions, which equates to around 358 sorties and 83 munitions per month if distributed evenly across all 12 months.[54] By the following June in 2016, the Air Force reported conducting 9,100 UCAV sorties and expending 3,400 munitions in 1,800 strikes since the campaign had begun in August 2014, or roughly 413 sorties and 154 munitions expended per month if allocated evenly across all 22 months.[55] Put another way, in August 2015, one year into the war, sorties by Air Force UCAVs counted for around 14 per cent of all crewed and uncrewed combat and ISR sorties in OIR and 4 per cent of the munitions expended. Seven months later, Air Force UCAVs represented 27 per cent of all combat and ISR sorties and 7 per cent of munitions expended in the war by that point. Other signs point to growing UCAV support for OIR in this period. Between 2015 and 2016, the US broke ground on a new UCAV facility at Jordan's Muwaffaq Salti Air Base, adding 12 large clamshell shelters and aprons for drones, as well as at a separate operating facility at Jordan's H4 Air Base.[56] The US also expanded UCAV infrastructure at Ali Al Salem Air Base in Kuwait to accommodate more Reapers and Predators.

The increase in UCAV activity between 2015 and 2016 corresponds with the overall progress of the campaign and reflects the apex of Coalition operations in OIR – the battles for Mosul in Iraq and Raqqa in Syria in 2016 and 2017, respectively. Depending on the operation, the proportion

of UCAV strikes relative to those by crewed aircraft fluctuated, with UCAVs tending to assume a larger role in battles for urban centres than they appear to have done throughout the entirety of Inherent Resolve. In the four-month-long battle for Manbij in northern Syria, US MQ-1Bs and MQ-9s conducted 500 sorties, accumulating 11,000 hours and launching 300 Hellfire missiles against ISIS targets, approximately 40 per cent of the munitions expended by Coalition aircraft.[57] Likewise, in the nearly five-month-long Battle of Raqqa in Syria in 2017, US Air Force MQ-1 Predator and MQ-9 Reapers were reportedly responsible for 20 per cent of all Coalition airstrikes, flying more than 44,000 hours in support of the operation.[58] Air Force UCAVs also contributed indirectly to strikes by crewed aircraft. For every two munitions launched by a UCAV in the first year of the war, an Air Force Predator or Reaper conducted roughly one 'buddy lase,' using a laser to guide a weapon from another aircraft to the target.[59] Some units conducted more assists for crewed aircraft than strikes; one Air Force attack squadron reported buddy lasing 196 laser-guided munitions to targets and conducting 170 independent Hellfire strikes in 2015.[60]

The pace of US air operations in support of Inherent Resolve declined appreciably after the Coalition's defeat of ISIS in Raqqa in September 2017. Still, a report by US investigators in spring 2021 warned that the threat from ISIS persists and that the group is using the Syrian Desert as a refuge to regroup.[61] In March 2021, Coalition forces conducted 133 airstrikes against ISIS in northern Iraq, the most in two years. Having defeated ISIS strongholds in Iraq and Syria, the Coalition is currently in what it calls the 'nor-

An MQ-1C Gray Eagle unmanned aircraft makes its way down an airfield on Camp Taji, Iraq, before a surveillance mission in the Baghdad area. (USAF/SGT Roland Hale)

A US Army MQ-1C Gray Eagle at Al Asad Air Base in Iraq in 2017. (US Army/CPT Stephen James)

malize' phase, or Phase IV, of the campaign plan, in which it seeks to build the capacity of local forces.

One of the most politically significant events involving drones in Iraq occurred long after the peak of Coalition operations against ISIS and had little directly to do with the campaign. In January 2020, a US drone strike outside of Baghdad killed Iran's Gen Qassem Soleimani, the leader of the Quds Force of the Islamic Revolutionary Guard Corps.[62] According to the *New York Times*, US MQ-9 Reapers targeted a convoy carrying Soleimani from the Baghdad airport to the city. The strike further escalated tensions between Washington and Tehran that had been rising since the Trump administration's withdrawal from the multinational nuclear agreement with Iran. Days later, Iran launched a retaliatory missile strike at Al Asad Air Base, injuring dozens of personnel.[63] After Iran accidently downed a civilian airliner over its territory in the mistaken belief that it was a US missile, tensions de-escalated between the two countries.

The Soleimani strike marked the first time that US drones had targeted a member of a foreign military with which it was not formally at war, a break from the past two decades of targeted killings of leaders of militant organisations. The Trump administration defended the necessity of the strike, arguing that Soleimani's efforts to galvanize Iraqi Shiite militias to attack Americans in Iraq represented an imminent threat. In a report in July, however, a U.N. official disputed this assessment, saying that the US lacked sufficient evidence of an imminent threat.[64] Despite Soleimani's death, the threat from Iran-backed Shiite militias in Iraq to US personnel and, more critically, to the authority of the Iraqi government has not evaporated. In February 2021, the US conducted airstrikes against an Iranian militias base in eastern Syria after a US contractor was killed in a rocket attack by militias.[65]

Libya (2011–today)

Data

Aircraft: Bayraktar-TB2, MQ-1 Predator, MQ-9 Reaper, Wing Loong 1, Wing Loong 2

Armament: AGM-114 Hellfire, Blue Arrow-7, GBU-12, MAM-C, MAM-L

Turkey/Government of National Accord

In 2019, Turkey began sending military advisers, proxy militias, and Bayraktar-TB2 UCAVs to Libya to support of the Government of National Accord (GNA), the U.N.-recognized government in Tripoli that took power in the years following the ouster of Libyan dictator Muanmmar Gaddafi. At the time, the GNA was struggling to defend itself against attacks by the Libyan National Army, an eastern Libyan group led by Khalifa Haftar that, with the support of the UAE and other partners, aimed to wrest power from the GNA. Ankara's motivations for supporting the GNA were multifaceted, writes Jalel Harchoui in a 2020 paper for the Foreign Policy Research Institute, and likely reflected a combination of ideological support for Islamist parties in the GNA, a political desire to assert nationalist Turkish prerogatives in the region, and an economic incentive in the form of long-standing Turkish commercial interests in Libya.[66]

In early 2019, Gen Khalifa Haftar and the Libyan National Army (LNA) launched a military offensive on Libya's capital, Tripoli, imperilling the GNA. At some point in spring 2019, Turkish Bayraktar-TB2s began conducting defensive airstrikes to prevent the LNA from taking the city. Shortly thereafter, LNA forces reporting shooting down several Bayraktar-TB2s, the first of some 15 Turkish UCAVs claimed by the LNA over between May and October 2019.[67] According to a 2019 report by a U.N. panel of experts, Turkish UCAVs were initially susceptible to the Russian-made Pantsir S-1 air defence system operated by mercenaries on behalf of the LNA. And until the GNA emplaced ground relay stations in late 2019, the Bayraktar-TB2s range also limited the types of operations in which it was able to engage.[68] Still, by early 2020, Turkish military support in the form of UCAVs and other ISR capabilities to the GNA helped stabilize its position in Tripoli, eliminating the air superiority of Haftar's forces and allowing the GNA to launch a spring counter-offensive aimed at regaining territory in western Libya.

It was during this period that the LNA lost several Pantsir S-1 air defence systems to Turkish UCAV strikes, a stark change from the Pantsir's success in downing Bayraktar-TB2s in the prior year. Observers have attributed the Pantsir's reversal in fortunes to a number of factors, including the inexperience of mercenary and Libyan crews, Turkey's use of Koral electronic warfare systems to jam the Pantsir's radar, and the vulnerability of the Pantsir to multi-directional air attacks as LNA forces retreated eastward. In a 2021 report, the U.N. panel of experts concludes that the 'introduction by Turkey of advanced military technology into the conflict was a decisive element in the often unseen, and certainly uneven, war of attrition that resulted in the defeat of [the LNA] in western Libya during 2020.' In addition to the Bayraktar-TB2s and Koral system, the U.N. panel also documented the use of Turkish Kargu-2 loitering munitions in Libya.

United Arab Emirates/Libyan National Army

The UAE's military engagement in Libya began in 2011, when it supported the NATO-led campaign against Libyan dictator Muanmmar Gaddafi. Beginning in 2013, the UAE deepened its involvement in Libya, providing military support to General Khalifa Haftar's campaign against Islamist militant groups in eastern Libya. Haftar's successes in securing eastern Libya prompted him to launch a campaign for control of the country, pitting his self-styled Libyan National Army (LNA) against the Government of National Accord (GNA), the U.N.-recognized government in Tripoli. The UAE, which views the GNA as sympathetic to Islamist groups like the Muslim Brotherhood, has continued to back Haftar in his bid for control, as has Egypt, France, and Russia, among others. In doing so, the UAE provided the LNA with various forms of material support, including Chinese-made strike-capable drones.

According a 2017 report by the UN Panel of Experts on Libya, the UAE has deployed Wing Loong I UCAVs to Libya since at least June 2016. The report found that the UAE was expanding facilities at Al Khadim Air Base in northeast Libya to accommodate Wing Loong Is and other, crewed combat aircraft.69 In a 2019 report, the U.N. panel concluded that the UAE had supplied Wing Loong IIs to its partners in Libya beginning in at least 2019.70 The U.N. panel found that four airstrikes in April 2019 involved a

Blue Arrow BA-7 air-to-surface missile, the same one that is used by the Wing Loong-series UCAVs. The Wing Loong-series UCAVs are the only platform involved in the conflict capable of equipping with the BA-7. Other visual imagery from Libya including video footage and satellite photographs cited in the U.N. reports further implicated the Wing Loong IIs – and, by extension, the UAE, the only confirmed operator of the aircraft in the region as of this writing – in the conflict.

Emirati drone activity in Libya appears to have peaked in 2019 and 2020, propelling the LNA to the outskirts of Tripoli in an attempt to take Libya's capitol.[71] It was during this period that Emirati-operated Wing Loong IIs were alleged to have conducted strikes that killed non-combatants. According to a report by Human Rights Watch, an Emirati drone strike in November 2019 on a factory in Wadi, al-Rabie, south of Tripoli, killed eight civilians.[72] Meanwhile, an independent investigation by the BBC found that an Emirati-operated Wing Loong II was responsible for an airstrike at a military academy in Tripoli in January 2020 that killed 32 unarmed cadets.[73] The US State Department, citing reports by human rights organisations and media outlets, said that strikes by Emirati drones and crewed aircraft in Libya have resulted in more than 130 civilian casualties, though it is unclear what proportion of these can be attributed to drones.[74]

United States

In March 2011, the US joined Operation Unified Protector, the seven-month-long NATO-led operation to enforce U.N. sanctions on the government of Libyan dictator Muanmmar Gaddafi. In April 2011, President Obama approved the use of armed MQ-1B Predators in support of Unified Protector.[75] By all accounts, the Predators played an outsized role in Unified Protector as an ISR and strike asset, achieving a number of firsts. A US MQ-1 Predator struck a Libyan 9K33 Osa (ASCC SA-8 Gecko) surface-to-air missile system in April 2011 in what was possibly the first time that a Predator conducted a destruction of enemy air defences mission.[76] During Unified Protector, US Predators conducted the first 'buddy lase' for strikes by a helicopter operated by a foreign nation, in addition to those for foreign fighter aircraft.[77] By August, Reuters reported that the US Predators had conducted 101 AGM-114 Hellfire strikes in the first five months of Unified Protector.[78] In October 2011, a Predator joined French Air Force Mirage 2000 jets in striking a convoy carrying Muanmmar Gaddafi, who died shortly thereafter.[79] The US deployed up to 10 MQ-1B Predators from Sigonella Air Base in Sicily, according to researchers at the RAND Corporation.

By the end of Unified Protector, the US reported conducting 450 UAV missions by MQ-1B Predators and RQ-4 Global Hawks, though it did not specify how many of these involved Predator strikes.[80] Separately, NATO reported conducting a total of 509 UAV missions over the course of the campaign – a figure which presumably includes the limited number of sorties by unarmed French Harfang and Italian MQ-9 Reaper UAVs in addition to those by US Predators – of which around 250 were strike missions.[81] Lt Col Gary Peppers, commander of 324th Expeditionary Squadron during Unified Protector, later said that the Predators shot 243 Hellfire missiles in 241 strikes over the course of the campaign. Remarkably, Predator strikes in Unified Protector were responsible for 'over 20 per cent of the total of all Hellfires expended in the 14 years of the [Predator's] deployment,' according to Peppers.[82] Despite a slight variation, these figures suggest that around half of the US UAV missions involved a strike.

Following Gaddafi's death and the conclusion of Operation Unified Protector, the US continued use unarmed UAVs to gather intelligence over Libya amid rising political instability and militia violence in the country. During the September 2012 attack on the US diplomatic mission in the eastern city of Benghazi, in which Islamist militants killed Ambassador J. Christopher Stevens and three other Americans, an unarmed US Predator drone provided intelligence support to US decision makers.[83] According to an investigation by the New America Foundation and Airwars, between 2012 and 2015, the US conducted a fewer than a dozen airstrikes in Libya, though it is unclear if these were by UCAVs or inhabited aircraft.[84] In August 2016, the US launched Operation Odyssey Lightning, a joint operation with Libya's Government of National Accord to retake the city of Sirte from a local affiliate of the self-styled Islamic State of Iraq and Syria (ISIS). Of the 495 US airstrikes in Odyssey Lightning, around 60 per cent, or nearly 300, were conducted by three Air Force MQ-9 Reapers, with the remainder by inhabited Marine Corps combat aircraft.[85] According to the Air Force, Odyssey Lightning served to validate new tactics for employing UCAVs, with Reapers often working in close coordination with each other to identify and engage targets and provide close air support to partner forces on the ground.

Between the conclusion of the Odyssey Lightning in early 2017 and today, intermittent US drone strikes have continued to target members of ISIS, Al Qaeda, and other extremist groups in Libya. In 2018, a US drone strike killed recruiter and logistician for Al Qaeda in southern Libya.[86] Then, in 2019, US MQ-9 Reapers based in neighbouring Niger conducted several more strikes in southern Libya targeting ISIS members.[87] In December 2019, Russian mercenaries fighting for the Libyan National Army downed a US Reaper near Tripoli.[88]

A Blue Arrow missile from a downed Wing Loong in 2019. (via Social Media)

Two Bayraktar-TB2s at Misrata air base in Libya in June 2020. (GoogleEarth)

Mali and Sahel (2018*–today)

Data
Aircraft: MQ-9 Reaper
Munitions: GBU-12

France

In 2013, France deployed unarmed Harfang UAVs to Niger to support Operation Serval, a counterterrorism operation in Mali against Islamist militants. The following January in 2014, France added newly acquired unarmed MQ-9 Reapers to support Serval. In mid-2014, France concluded Operation Serval and launched Operation Barkhane, an anti-insurgency campaign that, while focusing on Mali, encompasses several countries in the Sahel and remains ongoing as of this writing. France armed its Reapers beginning in 2019 and carried out its first drone strike Mali in December 2019.[89] As of spring 2021, French MQ-9 Reapers had accumulated more than 40,000 flight hours in the Sahel over the course of the previous six years.[90] In 2020, French Reapers conducted 590 sorties and accumulated 7,552 flight hours, exceeding those conducted by crewed strike and attack aircraft in support of Barkhane, according to statistics released by the Ministry of Defence.[91] According to French military officials in testimony before the National Assembly in October 2020, Reapers were directly responsible of 40 of the 97 airstrikes – 41 per cent – conducted by French forces between January and October 2020.[92] An April 2021 committee hearing by the National Assembly put Reaper strikes at approximately 45 per cent of all French airstrikes in Operation Barkhane in 2020.[93]

France's UCAV operations in Mali marked the first time that it has deployed armed drones. As of this writing, France has one system comprising of three MQ-9 Reaper aircraft deployed to Niamey, Niger in support of Barkhane. Given the expansive Sahel theater of operations and limited numbers of available crewed combat aircraft, the deployment fulfilled a long-held desire by the French military. It joined the U.S., Italy, and the U.K. in operating armed MQ-9 Reapers. The French Air Force's 1/33 Squadron 'Belfort' were responsible for Reaper operations until spring 2021,

A French Air Force MQ-9 Reaper at Diori Hamani International Airport, Niamey, Niger. (Armée de l'Air)

when they were replaced by the newly created 2/33 Squadron 'Savoie'.[94] The French Reapers are armed with GBU-12 guided bombs, though France intends to expand its equipage to include Hellfire missiles and GBU-49 bombs by the end of 2021. French Reapers have supported airstrikes by crewed fighter aircraft by identifying and lasing targets.[95] Some of these strikes have been criticized for killing or injuring civilians. A U.N. investigation found that a January 2021 French airstrike involving an MQ-9 Reaper mistakenly killed 19 civilians who were attending a wedding, a claim Paris has denied.[96]

United States

In the year following the 2012 Tuareg rebelling in Mali, the U.S. deployed unarmed MQ-1 Predators to neighbouring Niger. Initially, the deployment sought to provide intelligence and targeting support for French counter-terrorism operations in Mali. The U.S. drone presence in Niger would eventually expand to include U.S. MQ-9 Reapers and MQ-1C Gray Eagles, as well as other its participation in other missions such as countering transnational crime and trafficking in the region. In 2017, the Nigerien government permitted the U.S. to deploy armed drones on its territory and the U.S. began doing so the following year, according to press reports.[97] In 2019, U.S. Africa Command announced that a new, dedicated drone base outside of Agadez, a city in northern Niger, was fully operational.[98] Although the U.S. continues to conduct UCAV missions in the Sahel – in early 2021, an MQ-1C Gray Eagle armed with Hellfire missiles malfunctioned and crashed in Niger – the extent to which the U.S. has conducted drone strikes is unclear.[99] It is evident, however, that the U.S. contribution to the French-led counterterrorism mission in Mali is significant. According to French officials, around 40 percent of ISR missions in support of Operation Barkhane in 2020 were carried out by U.S. aircraft, likely largely by drones.[100]

Nagorno-Karabakh (2020)

Data
Aircraft: Bayraktar-TB2
Munitions: MAM-C, MAM-L

Azerbaijan

Azerbaijan has employed UCAVs and loitering munitions at various points in recent armed conflicts with Armenia over Nagorno-Karabakh, a disputed territory in southwestern Azerbaijan that is claimed by both countries. When violence flared in the territory during the Four Day War in 2016, video footage showed Azerbaijani Harop loitering munitions conducting strikes on Armenian positions.[101] However, it was during the 2020 Nagorno Karabakh War, or Second Nagorno Karabakh War as it is commonly known, that Azerbaijan wielded UCAVs and loitering munitions to particular effect. The 2020 conflict underscored how the diffusion of UCAV systems and technologies can create asymmetric advantages over opponents that, once exploited, can lead to tactical and operational gains. The conflict also indicated that UCAVs could provide an alternative to inhabited combat aircraft against adversaries equipped with conventional – if aging – air defences, reducing the threats to valuable aircrew and aircraft. Azerbaijani UAVs and UCAVs were based at Kyurdamir Air Base and Yevlakh Airport.

The Second Nagorno-Karabakh War confirmed Azerbaijan's acquisition of Bayraktar-TB2s, becoming the fourth country to operate the Turkish UCAVs. The Bayraktar-TB2s were Azerbaijan's first UCAV acquisition and joined a fleet of Israeli-made UAVs for ISR and loitering munitions. On September 27, the day war resumed in Nagorno-Karabakh, Azerbaijan's Ministry of Defence posted six videos to its official YouTube channel featuring aerial footage that contained visual characteristics typical of video footage from a Bayraktar-TB2. In an October 5 interview with a Turkish pro-government television channel, Azerbaijani President Ilham Aliyev attributed Azerbaijan's battlefield success to the fact that it was operating 'advanced Turkish drones'. The strongest evidence of Bayraktar-TB2s in Nagorno-

The launch of an Azerbaijani Harop loitering munition. (Azeri News Agency)

Karabakh came on October 19, when the official YouTube Artsakh Defence Forces YouTube channel posted a video of the remnants of what appeared to be a downed Bayraktar-TB2. Photos published to social media on October 24 by independent journalists also showed pieces of a Bayraktar-TB2.

Azerbaijan's decision to shift to a drone-centric air campaign occurred in the early days of the war and was likely motivated by the challenging mountainous topography of Nagorno-Karabakh and Azerbaijani setbacks on the ground. Armenian air defences were the initial priority of Azerbaijan air attacks. Azerbaijan reportedly used uninhabited Soviet-era An-2 biplanes to serve as decoys, exposing Armenian air defences and distracting them from the main attack, a tactic that has been used repeatedly by Israel and others from the earliest days of modern drones.[102] To target Armenian air defences, Azerbaijan used a combination of UCAVs and loitering munitions like the Harop, which can be equipped with an antiradiation sensor to automatically target adversary radars. At Royal United Services Institute, Jack Watling and Sidharth Kaushal write that loitering munitions are difficult for many radars to detect and are a cost-effective solution to defeating expensive air defence systems like the Russian-made S-300.[103] The Turkish Bayraktar-TB2s also appeared be successful against the air defence systems; in an interview with the *Daily Beast*, an Artsakh defence official said that the Turkish drones flew too high for the Armenian air defences to intercept.[104] The destruction of Armenia's air defences cleared the way for a wider air campaign against armoured vehicles and Armenian positions, resulting in an Azerbaijani ground offensive north from the lowlands below Nagorno-Karabakh. At the European Council on Foreign Relations, Gustav Gressel argues that Azerbaijani drones were particularly effective in bringing firepower against Armenia's reserve forces, either through direct attacks or by spotting targets for Azerbaijani artillery and other long-range weapons, reducing Armenia's ability to reinforce and resupply forward positions in Nagorno-Karabakh.[105] This may have had a greater effect on the outcome of the war, argues Gressel, than the alleged – and probably exaggerated – sums of Armenian assets destroyed in strikes by Azerbaijani drones. And whatever their success may have been in targeting air defences and vehicles, Azerbaijan's drones did not alleviate the realities of taking and holding territory from an enemy in entrenched positions in mountainous terrain. Publicly

Azerbaijan exhibits the Bayraktar TB2 in the Baku Victory Parade in December 2020. (Presidential Press and Information Office of Azerbaijan)

available estimates of combat deaths on both sides in the war are roughly similar – around 3,360 Armenians and 2,820 Azerbaijanis – despite Azerbaijan's qualitative advantage in airpower.[106]

Several factors appear to have contributed to Azerbaijan's successful use of UCAVs and loitering munitions in the Second Nagorno-Karabakh War.

First, Azerbaijani drones faced no opposition from Armenia's Su-30 interceptors, possibly a consequence of Turkey's deployment of F-16 fighters to Azerbaijan and the ensuing risk of a direct confrontation between Armenia and Turkey.

Second, Armenia's deficit in electronic warfare and specialized counter-UAV systems contributed to its inability to defend against drones. Armenia's reliance on Soviet-era equipment and preference for surface-to-air missiles for air defence was inadequate, particularly when faced with saturation attacks from drones.[107] However, multiple unverified accounts have emerged regarding the use of electronic warfare systems by both sides. Russian media has claimed that its deployment of the Krasukha-4 EWS towards the end of the conflict helped down several Bayraktar-TB2s.[108] Meanwhile, some observers have suggested that Azerbaijani UCAV strikes on Armenian air defences may have been aided by electronic warfare systems, perhaps the Turkish KORAl EW system that helped Bayraktar-TB2s destroy Russian-made air defence systems in Libya and Syria.[109] Neither accounts have been confirmed, though both suggest that offensive and defensive electronic warfare systems are likely to assume greater prominence in future conflicts as drones proliferate.

Third, Turkish military personnel appear to have had a direct role in piloting Bayraktar-TB2s and are widely believed to have helped Azerbaijan plan its offensive. (It was not until February 2021, months after the campaign concluded, that Azerbaijani military personnel completed their four-month-long training course on operating and maintaining the Bayraktar-TB2.[110]) Just as they had in Turkey's intervention in Libya, Turkish specialists and their commensurate experience from years of UCAV operations in other theatres likely contributed to Azerbaijan's successes.

Nigeria (2016–today)  

Data
Aircraft: CH-3B
Munitions: AR-1

Nigeria

In February 2016, the Nigerian government published video footage that appeared to show an airstrike by a Nigerian-operated Chinese-made CASC CH-3B UCAV.[111] The strike reportedly targeted a base belonging to Boko Haram, an extremist Islamist group that operates in north-eastern Nigeria and in neighbouring countries. In the years that followed, however, the CH-3s appeared to make a negligent contribution to Nigeria's efforts to combat the group. In testimony before the US Senate Armed Services Command in April 2019, General Stephen Townsend attributed Nigeria's apparently infrequent use of CH-3s to the 'poor quality' of Chinese equipment.[112] Air Marshal Sadique Abubakar, then the head of the Nigerian Air Force, explained in April 2019 that the high cost of maintaining UAVs posed a challenge to the service.[113] Nonetheless, in late 2020, Nigeria announced that it would acquire eight new Chinese-made UCAVs, including two AVIC Wing Loong-2s, a significant upgrade to its fleet and one that is likely to result in a larger role for Nigeria's UCAVs in its fight against Boko Haram.[114]

Photos of the Wing Loong II and the ground control station reportedly on order with the Nigerian Air Force. (NAF)

Somalia (2007–today) 

Data
Aircraft: MQ-9 Reaper
Munitions: AGM-114 Hellfire, GBU-12

United States

In the years following the 11 September 2001, attacks, the US conducted intermittent airstrikes and special operations raids on targets associated with Al Shabab, the Islamist militant group allied to Al Qaeda. The first US drone strike occurred on 11 June 11 2011 and targeted an Al Shabab training camp south of the city of Kismayo.[115] Over the next five years, the US conducted a limited number of drone strikes in Somalia. According to the Bureau of Investigative Journalism, a London-based non-governmental organisation, the US is believed to have conducted 28 air operations in Somalia in this period, of which 20 were likely drone strikes.[116] Beginning in 2017, the pace of US airstrikes in Somalia accelerated, doubling between 2016 and 2017. Observers have attributed this increase in activity to the Trump administration's decision to weaken the Obama administration's targeting rules for counterterrorism operations by giving US military commanders greater flexibility to waive certain requirements designed to avoid civilian casualties.[117] Strikes appear to have peaked in 2019, when the US reportedly conducted nearly 60 airstrikes, although it is unclear precisely how many of these were by drones. The US also began targeting members of a local affiliate of the self-styled Islamic State of Iraq and Syria (ISIS) in 2019.[118]

Syria (2014–today)

Data

Aircraft: Bayraktar-TB2, MQ-1B Predator, MQ-9A Reaper, Orion, Shahed-129

Munitions: AGM-114 Hellfire, GBU-12, MAM-C, MAM-L

Iran

Since the start of the Syrian Civil War in 2011, Iran has supplied the Syrian regime of Bashar al-Assad with weapons and personnel, including with various types of drones. In 2015, Iranian media outlets published video footage of what appeared to be a Shahed-129 conducting strikes in Syria.[119] In a speech the following year, Iranian general Mohaamad Bagheri announced that Iranian UAVs were 'being used to hit terrorist targets in Iraq and Syria.'[120] Since then, however, there are few recorded instances of Iranian UCAVs engaging in strikes Syria. Iran's UCAVs were featured in two of the handful of air-to-air engagements of the conflict. In early June 2017, an Iranian Shahed-129 launched a missile at a vehicle near the al-Tanf garrison, which housed US-backed forces in south-eastern Syria. Shortly thereafter, a US Air Force F-15E Strike Eagle shot down the Shahed-129 as it attempted to return to the area. A few weeks later, a US fighter downed a second Shahed-129.[121] The two incidents are among the relatively small number of air-to-air engagements involving drones in any conflict to date. Separately, in 2018, Israel reported downing an armed Iranian drone that took off from Syria.[122]

Russia

In November 2019, Russian media outlets reported that the Kronstadt Orion had arrived in Syria for operational trials.[123] Then, in February 2021, a Russian television broadcast suggested that the Orion had carried out at least 17 strike missions in Syria. The video showed footage of the Orion launching several unidentified missiles and a 100kg (220lb) bomb.[124] Subsequent Russian media reports highlight the role that the KRUS Strelets, an infantry-portable battlefield C4I system that is commonly teamed with smaller Russian UAVs for target acquisition, played in finding and fixing targets for the Orion.[125] The Orion in the video was based at Tiyas Military Air Base (T-4), an Assad regime and Russian facility in central Syria. Although the Orion did not appear to have much of a battlespace presence in Syria and it remains unclear when precisely these strikes occurred, they appear to be Russia's first UCAV operations.

Turkey

Beginning in 2016, Turkish UAVs have participated in multiple campaigns in northern and western Syria. Turkish Bayraktar-TB2s are believed to have played a limited role in Operation Euphrates Shield (Fırat Kalkanı Harekâtı), a cross-border campaign against Syrian Kurdish organisations in 2016 and 2017. The precise role of armed UCAVs in this conflict remains somewhat unclear; poor weather may have hampered air operations and at least three-quarters of Bayraktar-TB2s in the Turkish inventory at the time were unarmed and used for intelligence and surveillance and for target acquisition for crewed combat aircraft.[126]

Turkey's first significant deployment of UCAVs occurred in 2018 in Operation Olive Branch (Zeytin Dalı Harekâtı), a four-month-long Turkish military operation against Syrian Kurdish groups around the district of Afrin. In public statements, Selcuk Bayraktar, an executive at Baykar Defence, has claimed that Bayraktar-TB2 missions represented over 90 per cent of all Turkish military sorties in Operation Olive Branch, an oft-repeated statistic in Turkish media that is difficult to verify.[127] The Turkish military attributes around 11 per cent of Kurdish militants killed in the conflict to strikes by Bayraktar-TB2 UCAVs between January and April 2018, a figure that rises to 28 per cent if one includes targets marked by the TB2 for strikes by crewed aircraft.[128] Regardless of the precise figures, Operation Olive Branch was Turkey's first wide-scale deployment of UAVs and UCAVs, marking the beginning of a deeper integration of UCAVs into Turkish air and ground operations and setting the tone for successive military interventions. Speaking at a conference in March 2018, Turkish Prime Minister Binali Yildirim praised the role that Bayraktar-TB2s played in Olive Branch, arguing that they 'altered the fate' of the operation.[129] Operation Olive Branch also reflected a larger role for airpower at the outset of the campaign to achieve operational objectives than it had played in prior cross-border incursions. Turkish analyst Can Kasapoglu has argued that the high air sortie rates at the start of Olive Shield reflected lessons learned from Euphrates Shield, in which artillery fires played a greater role; namely the need to use air-launched precision-guided munitions to target the adversary's subterranean capabilities and maximise the overall psychological effects of airpower to improve the survivability of land power forces.[130]

If Operation Olive Branch signalled a Turkey's embrace of drones and airpower, Operation Spring Shield, a February to March 2020 counteroffensive against the Syrian government in north-western Syria's Idlib Province, highlighted new developments in Turkey's combat applications of UCAVs.[131] Following deadly attacks by Russian and Syrian forces on Turkish soldiers in late February, Turkey launched retaliatory strikes aimed at preventing a Syrian ground offensive into Idlib Province. These strikes appear to have been largely carried out by Anka-S and Bayraktar-TB2 UCAVs. Gun camera footage reportedly from drones appeared to show dozens of tanks, armoured vehicles, and artillery and air defence systems destroyed in airstrikes, though it is difficult to know with certainty the extent of the damage caused by Turkish drones.[132] Unlike the relatively permissive environments of prior Turkish interventions into northern Syria, in Spring Shield Turkey had to contend with Russian-made air defence systems and the Syrian Arab Air Force's warplanes. According to widely disseminated accounts in Turkish media, the Turkish military relied on the Aselsan KORAL electronic warfare system to jam and deceive Russian-made air defences, clearing the way for strikes from Turkish drones and jets.[133] After Syrian fighter jets downed a Turkish drone, Turkish F-16s intercepted Syrian warplanes over Idlib and bombed Nayrab airport, west of Aleppo city in northern Syria.[134]

In Operation Spring Shield, Turkish UCAVs – primarily the Bayraktar-TB2 – were the centrepiece of both an intensive air-centric campaign for Idlib and an information offensive. Aerial footage apparently taken by drones of strikes on Syrian military hardware dominated the airwaves of Turkish news media outlets and the social media feeds of government officials.[135] [136] In a televised ceremony on 3 March 2020 at Hatay airport in southeast Turkey, the primary base for Bayraktar-TB2s operating over Idlib, İsmail Demir, president of the Turkish arms directorate (SSB), and Baykar Defence official Selçuk Bayraktar celebrated the contribution that Turkish military technology was making to the campaign.[137] These sentiments were echoed by government-friendly media outlets, some of which focused on Selçuk Bayraktar's prominent role. In a 3 March 2020 op-ed in the pro-government Hurriyet newspaper, one columnist expounded that *'Behind* [Turkey's drones] *are one company, one family and one young businessman.'*[138]

Although the prominence of Selçuk Bayraktar and of Bayraktar-TB2 in Turkish government and media narratives was not a new development – in the midst of the Olive Branch in 2018 Turkish President Tayyip Erdoğan visited a drone base where he autographed a Bayraktar-TB2[139] – it reached new heights in Spring Shield, becoming a defining feature of successive Turkish military engagements.

United Kingdom

In 2014, the UK Royal Air Force redeployed MQ-9 Reapers with the No. 13 Squadron from Afghanistan to Ali Al Salem Air Base in Kuwait to support Operation Shader, the UK's designation for the counter-ISIS campaign in Iraq and Syria. The UK began flying Reaper operations over Syria in October 2014.[140] In August 2015, an RAF Reaper carried out an airstrike in Syria that killed Reyaad Khan, a UK citizen who had joined ISIS.[141] According to data released by the UK government, UK MQ-9 Reapers had conducted 133 strikes in Syria between 2014 and 2020, roughly 25 per cent of all British airstrikes in Syria.[142] Strikes by RAF Reapers declined year-over-year from a peak of 66 in 2018 to 0 in 2020, which corresponds with the apex of the counter-ISIS campaign in Syria and the subsequent drop in the overall number of Reaper sorties in Syria. The RAF Reapers were armed with GBU-12 laser-guided bombs and AGM-114 Hellfire – primarily the AGM-114R2 variant – air-to-surface missiles.

United States (See Iraq)

Turkey (2016–today)

Data
Aircraft: Bayraktar-TB2
Munitions: MAM-C, MAM-L

Turkey

Turkey recorded its first drone strike in September 2016 in air operations in Çukurca, a province in southwestern Turkey.[143] In a statement, the Turkish military announced that an armed Bayraktar-TB2 killed five people associated with the Kurdistan Workers' Party (PKK), a Kurdish militant group, in the strike. Since then, Turkish UCAVs have been active in operations against Kurdish groups in Turkey, a conflict that frequently spills over into neighbouring Iraq.

A Bayraktar TB2 armed with MAM-L and MAM-C missiles. (TAI)

Yemen (2001–today)

Data
Aircraft: CH-4, MQ-1B Predator, MQ-9A Reaper, Wing Loong-1
Munitions: AGM-114 Hellfire, GBU-12

United States

In November 2002, a US MQ-1 Predator drone operated by the Central Intelligence Agency launched a strike that killed Qaed Salim al-Harethi, a leader of Al Qaeda in the Arab Peninsula (AQAP) and one of the alleged planners of the bombing of the USS *Cole* (DDG-67) in 2000, and five other Al Qaeda operatives.[144] Though not the first armed drone strike conducted by the United States, it marked the start of a years-long US effort to target leaders of the Al-Qaeda group and its allies outside of the territorial boundaries of conflict zones such as Afghanistan, one that would become colloquially known as the targeted killed campaign. In Yemen, however, the US did not resume strikes until some years later, after AQAP attempted to bomb a commercial airliner in December 2009.[145] In 2011, a US drone strike in Yemen killed the US-born cleric Anwar Al-Awlaki, an AQAP leader who was involved in planning the attempted attack in 2009.[146] In the years the followed, successive US strikes by the CIA and the Joint Special Operations Command (JSOC) would continue to target members of AQAP, killing other high profile members of AQAP.[147]

It is difficult to say how many of the US military actions in Yemen were carried out by drones as opposed to special operations forces or strikes by crewed combat aircraft. Three non-governmental organisations – the London-based Bureau of Investigative Journalism and the Washington-based New America Foundation and Long War Journal – have tracked US military actions in Yemen and elsewhere using largely public sources such as media reports. Given that these are unofficial accounts that are often not verified by the NGO and that there are discrepancies between the methodologies of each organisation, they should not be taken as an authoritative accounting of US actions. Nonetheless, they are relatively consistent with each other and can provide helpful context to a campaign that is has been conducted largely outside of the public eye. Of the 336 'confirmed' actions logged by the Bureau of Investigative Journalism between 2002 and 2020, around two-thirds have been tagged as an airstrike or drone strike.[148] US military

actions in Yemen peaked in 2017; in December, US Central Command announced that it had conducted over 120 strikes against AQAP and a local affiliate of the self-styled Islamic State, though it is unclear whether these were by drones or crewed aircraft. Since then, strikes attributed to the US appear to have declined sharply, as have independent efforts to track them.

Non-governmental organisations have criticised the US for conducting military actions – particularly drone strikes – in Yemen that have resulted in civilian deaths or injuries and for the lack of accountability for these incidents. Data from non-governmental organisations regarding non-combatant deaths varies; the Washington-based New America Foundation estimates between 115 and 149 civilians have been killed since 2002, while the Bureau of Investigative Journalism estimates between 174 and 224 deaths. Official statistics released by the Obama administration in 2016 are significantly lower than estimates by NGOs and do not specify how many of these casualties occurred in Yemen.[149] Experts have also disagreed over the legality of US drone strikes under domestic and international law, especially those strikes that have resulted in the deaths of US citizens like Anwar Al-Awlaki.[150] Survivors of US drone strikes in Yemen have challenged the legality of the strikes in courts in Europe and the US.[151]

United Arab Emirates

In 2015, the UAE joined a coalition led by Saudi Arabia to intervene in the Yemeni Civil War. Over the next several years, the UAE deployed several types of drones in support of coalition efforts against the Houthi group and Islamist militant groups, including Al Qaeda in the Arab Peninsula and the self-styled Islamic State. These included the strike-capable Chinese-made Wing Loong I and Wing Loong II and the unarmed US-made Predator XP. Emirati drones were based in Eritrea and Saudi Arabia. The extent of the UAE's drone strikes as a portion of the coalition's air campaign is unclear; there are a limited number of press reports that implicate the UAE in drone strikes. However, a 2018 report in *Foreign Policy* magazine found that the UAE was likely responsible for a drone strike that killed Saleh al-Samad, president of the Houthis' Supreme Political Council, in one of the most significant assassinations of the war.[152] The Emirati military withdrew from Yemen in 2020; in 2021, it began dismantling the base in Assab, Eritrea that had previously housed Chinese-made Wing Loong drones.[153]

15 Nauman, Q., 'Pakistan's Burraq Drone Kills Three Militants, Officials Say,' *Wall Street Journal*, 7 September 2015, https://www.wsj.com/articles/pakistans-burraq-drone-kills-three-militants-officials-say-1441632088

16 Dominguez, G., 'Pakistan receives five CH-4 UAVs from China,' *Janes*, 27 January 2021, https://www.janes.com/defence-news/news-detail/pakistan-receives-five-ch-4-uavs-from-china

17 United Kingdom, Ministry of Defence, Op HERRICK (Afghanistan) Aircraft Statistics, 29 October 2015, accessed at: https://www.gov.uk/government/statistics/operation-herrick-afghanistan-aircraft-statistics/

18 Whittle, R., 'How We Missed Mullah Omar,' *Politico*, 16 September 2014, https://www.politico.com/magazine/story/2014/09/how-we-missed-mullah-omar-111026

19 Tiernan, T., US Central Command Air Force, 'Reaper Drops First Precision-Guided Bomb, Protects Forces,' news release, 8 November 2007, https://www.af.mil/News/Article-Display/Article/125196/reaper-drops-first-precision-guided-bomb-protects-forces/

20 U.S. Air Forces Central Command Public Affairs, 'Combined Forces Air Component Command Airpower Statistics, news release, 6 January 2013, https://web.archive.org/web/20130307172213/http://www.afcent.af.mil/shared/media/document/AFD-130114-003.pdf

21 Molot, M., 'Bad Idea: Calling US Operations in Afghanistan, Iraq, and Syria 'Endless Wars', Defence360 (blog), Center for Strategic and International Studies, 7 January 2020, https://defence360.csis.org/bad-idea-calling-u-s-operations-in-afghanistan-iraq-and-syria-endless-wars/

22 Garland, C., 'Surveillance surge puts more Reapers at Afghan airfield than any other,' *Stars and Stripes*, 26 January 2018, https://www.stripes.com/news/surveillance-surge-puts-more-reapers-at-afghan-airfield-than-any-other-1.508593

23 Burgess, R. R., 'Marines to Operate Armed Reaper UAS in Coming Months,' *SeaPower*, 9 September 2020, https://seapowermagazine.org/marines-to-operate-armed-reaper-uas-in-coming-months/

24 Shephard News Team, 'Legal issue prompts USMC to buy contractor-operated Reapers,' *Shephard News*, 14 April 2021, https://www.shephardmedia.com/news/uv-online/legal-issue-prompts-usmc-buy-contractor-operated-r/

25 Frontline staff, 'Nek Muhammad,' *PBS Frontline*, 3 October 2006, https://www.pbs.org/wgbh/pages/frontline/taliban/militants/mohammed.html

26 Coll, S., 'The Unblinking Stare: The Drone War in Pakistan,' *The New Yorker*, 17 November 2014, https://www.newyorker.com/magazine/2014/11/24/unblinking-stare

27 Warrick, J., *The Triple Agent: The Al-Qaeda Mole Who Infiltrated The CIA*, p190

28 Schmitt, E., 'American Strike Is Said to Kill a Top Al Qaeda Leader,' *The New York Times*, 31 May 2010, https://www.nytimes.com/2010/06/01/world/asia/01qaeda.html

29 Warrick, J., *The Triple Agent: The Al-Qaeda Mole Who Infiltrated The CIA*, p202

30 New America staff, 'America's Counterterrorism Wars', New America Foundation, accessed on 1 September 2021, https://www.newamerica.org/international-security/reports/americas-counterterrorism-wars/methodology

31 Entous, A., Gorman, S. and Barnes, J. E., 'US Tightens Drone Rules,' *Wall Street Journal*, 4 November 2011, https://www.wsj.com/articles/SB100014240529702046219045770139826729738 36

32 Mehsud, I. T. and Khan, I., 'Pakistan Taliban Say Deputy Leader Was Killed in US Strike,' *The New York Times*, 13 February 2018, https://www.nytimes.com/2018/02/13/world/asia/pakistan-taliban-commander.html

33 Gettinger, D., 'The Disposition Matrix,' Center for the Study of the Drone at Bard College, 25 April 2015, https://dronecenter.bard.edu/the-disposition-matrix/

34 Wahab, B., 'Iran's Missile Attack in Iraqi Kurdistan Could Backfire,' Washington Institute for Near East Policy, 11 September 2018, https://www.washingtoninstitute.org/policy-analysis/irans-missile-attack-iraqi-kurdistan-could-backfire

35 MEMO Staff, 'Iran launches drone and missile strikes on Kurdish positions in Iraq,' *Middle East Monitor*, 14 July 2019, https://www.middleeastmonitor.com/20190714-iran-launches-drone-and-missile-strikes-on-kurdish-positions-in-iraq/

36 Coles, I. and Nissenbaum, D., 'US: Saudi Pipeline Attacks Originated From Iraq,' *The Wall Street Journal*, 28 June 2019, https://www.wsj.com/articles/u-s-saudi-pipeline-attacks-originated-from-iraq-11561741133/

37 Pamuk, H., 'Exclusive: US probe of Saudi oil attack shows it came from north – report,' *Reuters*, 19 December 2019, https://www.reuters.com/article/us-saudi-aramco-attacks-iran-exclusive/exclusive-u-s-probe-of-saudi-oil-attack-shows-it-came-from-north-report-idUSKBN1YN299/

38 MEE and agencies, 'Iraq drone attack targets US forces at Erbil airport,' *Middle East Eye*, 15 April 2021, https://www.middleeasteye.net/news/iraq-us-erbil-airport-drone-attack-forces-targeted

39 Office of Inspector General, Operation Inherent Resolve: Lead Inspector General Report to the United States Congress, Washington, D.C., GPO, 30 July 2021, https://www.dodig.mil/reports.html/Article/2716943/lead-inspector-general-for-operation-inherent-resolve-quarterly-report-to-the-u/

40 Stevenson, B., 'Iraq debuts new Chinese CH-4 UAV,' *FlightGlobal*, 13 October 2015,
https://www.flightglobal.com/civil-uavs/iraq-debuts-new-chinese-ch-4-uav/118492.article

41 Binnie, J., 'US-led coalition says only one Iraqi CH-4 UAV fully operable,' *Janes*, 7 August 2019,
https://www.janes.com/defence-news/news-detail/us-led-coalition-says-only-one-iraqi-ch-4-uav-fully-operable

42 Office of Inspector General, Operation Inherent Resolve: Lead Inspector General Report to the United States Congress, Washington, D.C., GPO, 30 July 2021,
https://www.dodig.mil/reports.html/Article/2716943/lead-inspector-general-for-operation-inherent-resolve-quarterly-report-to-the-u/

43 AFP staff, 'Turkey, Iran deploy drones in north Iraq against Kurd rebels,' *Agence-France Presse*, 2 October 2020,
https://www.arabnews.com/node/1743076/middle-east

44 McKernan, B., 'Kurds in 'mountain prison' cower as Turkey fights PKK with drones in Iraq,' *The Guardian*, 4 April 2021,
https://www.theguardian.com/world/2021/apr/04/iraq-turkey-pkk-drones-kurds-kurdistan

45 Majeed, R. and Abdulla, N., 'Turkish Drone Strike Leaves Civilian Casualties,' *Voice of America*, 2 July 2020,
https://www.voanews.com/extremism-watch/turkish-drone-strike-leaves-civilian-casualties

46 Al Jazeera staff, 'Iraq fumes against Turkey over deadly drone attack,' *Al Jazeera*, 12 August 2020,
https://www.aljazeera.com/news/2020/8/12/iraq-fumes-against-turkey-over-deadly-drone-attack

47 France 24 staff, 'Turkey, Iran deploy 'game-changing' drones in north Iraq,' *France24*, 10 January 2020, https://www.france24.com/en/20201001-turkey-iran-deploy-game-changing-drones-in-north-iraq

48 DroneWarsUK staff, 'British Drone Operations Against Isis, 2014–2016,' DroneWarsUK, February 2017,
https://dronewarsuk.files.wordpress.com/2017/02/uk-armed-drone-operations-against-isis-in-iraq-and-syria-feb2017.pdf

49 DroneWarsUK staff, 'UK Drone Strike Stats,' DroneWarsUK, last updated 27 April 2021, https://dronewars.net/uk-drone-strike-list-2/

50 Ibid

51 Clausen, C, 'Block 5 MQ-9 debuts in combat', 432nd Air Expeditionary Wing Public Affairs, 29 June 2017,
https://www.af.mil/News/Article-Display/Article/1232889/block-5-mq-9-debuts-in-combat/

52 Wasser et al., 'The Air War Against the Islamic State: The Role of Airpower in Operation Inherent Resolve', *RAND Corporation*, 2021, 303, https://www.rand.org/pubs/research_reports/RRA388-1.html

53 Gettinger, D, 'The Drone Databook', *Center for the Study of the Drone*, September 2019,
https://dronecenter.bard.edu/projects/drone-proliferation/databook/

54 Barclay, N., 'RPAs prove vital in fight against ISIS,' Creech Air Force Base, news release, 12 August 2015,
https://www.creech.af.mil/News/Article-Display/Article/669947/rpas-prove-vital-in-fight-against-isis/CreechAirForceBase/

55 Skowronski, W., 'Reapers: and the RPA Resurgence,' *Air Force Magazine*, August 2016,
https://www.airforcemag.com/PDF/MagazineArchive/Magazine%20Documents/2016/August%202016/0816reapers.pdf

56 'Secret U.S. Drone Base Reportedly Spotted in Jordan', *Haaretz*, 3 July 2017,
https://www.haaretz.com/us-news/secret-u-s-drone-base-reportedly-spotted-in-jordan-1.5491532

57 Clausen, C., 432d Wing Public Affairs, 'MQ-1, MQ-9 aircrews help liberate Manbij,' news release, 3 April 2017,
https://www.dvidshub.net/news/228951/mq-1-mq-9-aircrews-help-liberate-manbij

58 Everstine, B. W., 'Predators, Reapers Kept Persistent Watch as ISIS Lost Raqqa,' *Air Force Magazine*, 5 December 2017,
https://www.airforcemag.com/predators-reapers-kept-persistent-watch-as-isis-lost-raqqa/

59 Barclay, N., 'RPAs prove vital in fight against ISIS,' Creech Air Force Base, news release, 12 August 2015,
https://www.creech.af.mil/News/Article-Display/Article/669947/rpas-prove-vital-in-fight-against-isis/CreechAirForceBase/

60 Barclay, N., Creech Air Force Base, 'Guide me in: MQ-1s, MQ-9s provide 'buddy lase' capability against ISIL,' news release, 22 June 2016,
https://www.creech.af.mil/News/Article-Display/Article/810069/guide-me-in-mq-1s-mq-9s-provide-buddy-lase-capability-against-isil/CreechAirForceBase/

61 Office of Inspector General, 'Operation Inherent Resolve: Lead Inspector General Report to the United States Congress, Washington, D.C', GPO, 30 July 2021,
https://www.dodig.mil/reports.html/Article/2716943/lead-inspector-general-for-operation-inherent-resolve-quarterly-report-to-the-u/

62 Crowley, M., Hassan, F. and Schmitt, E., 'US Strike in Iraq Kills Qassim Suleimani, Commander of Iranian Forces,' *New York Times*, 2 January 2020, https://www.nytimes.com/2020/01/02/world/middleeast/qassem-soleimani-iraq-iran-attack.html

63 Qiblawi, T., Damon, A. and Laine, B., 'US troops sheltered in Saddam-era bunkers during Iran missile attack,' *CNN*, 14 January 2020,
https://www.cnn.com/2020/01/13/middleeast/iran-strike-al-asad-base-iraq-exclusive-intl/index.html

64 Nevehay, S., 'U.N. expert deems US drone strike on Iran's Soleimani an 'unlawful' killing,' *Reuters*, 6 July 2020,
https://www.reuters.com/article/us-usa-iran-un-rights/u-n-expert-deems-u-s-drone-strike-on-irans-soleimani-an-unlawful-killing-idUSKBN2472TW

65 GARAMONE, J., 'US Conducts Defensive Airstrikes Against Iranian-backed Militia in Syria,' Department of Defense, 25 February 2021, https://www.defence.gov/Explore/News/Article/Article/2516530/us-conducts-defensive-airstrikes-against-iranian-backed-militia-in-syria/

66 AP STAFF, 'Obama Oks use of armed drone aircraft in Libya,' *Associated Press*, 21 April 2011, https://www.cbsnews.com/news/obama-oks-use-of-armed-drone-aircraft-in-libya/

67 NATO, 'NATO delays airstrike, asks civilians to move away from military installations,' news release, 24 April 2011, https://www.nato.int/cps/en/natolive/news_72885.htm

68 PHINNEY, T. R., 'Reflections on Operation Unified Protector,' *Joint Forces Quarterly*, 73, Spring 2014, https://ndupress.ndu.edu/JFQ/Joint-Force-Quarterly-73/Article/577509/reflections-on-operation-unified-protector/

69 REUTERS STAFF, 'Factbox: US stepped up pace of Libya air strikes: Pentagon,' *Reuters*, 22 August 2011, https://www.reuters.com/article/us-libya-usa-strikes/factbox-u-s-stepped-up-pace-of-libya-air-strikes-pentagon-idUS-TRE77L76C20110822

70 ACKERMAN, S., 'Predator Teamed Up With French Jets to Stop Gadhafi,' *Wired*, 21 October 2011, https://www.wired.com/2011/10/predator-french-jet-kill-gadhafi/

71 MUELLER, K, 'Precision and Purpose: Airpower in the Libyan Civil War', *RAND Corporation*, 2015, https://www.rand.org/pubs/research_reports/RR676.html

72 NATO, https://www.nato.int/nato_static/assets/pdf/pdf_2012_05/20120514_120514-nato_1st_icil_response.pdf

73 TURSE, N., MOLTKE, H. and SPERI, A., 'Secret War: The US Has Conducted 550 Drone Strikes in Libya Since 2011 – More Than in Somalia, Yemen, or Pakistan,' *The Intercept*, 20 June 2018, https://theintercept.com/2018/06/20/libya-us-drone-strikes/

74 KAPHLE, A., 'Timeline: How the Benghazi attacks played out,' *Washington Post*, 17 June 2014, https://www.washingtonpost.com/world/national-security/timeline-how-the-benghazi-attack-played-out/2014/06/17/a5c34e90-f62c-11e3-a3a5-42be35962a52_story.html

75 BERGEN, P. and SIMS, A., 'Airstrikes and Civilian Casualties in Libya: Since the 2011 NATO Intervention,' New America Foundation, 20 June 2018, https://www.newamerica.org/international-security/reports/airstrikes-and-civilian-casualties-libya/

76 CLAUSEN, C., 'Providing freedom from terror: RPAs help reclaim Sirte,' 432nd Wing news release, 1 August 2017, https://www.acc.af.mil/News/Article-Display/Article/1265247/providing-freedom-from-terror-rpas-help-reclaim-sirte/

77 SCHMITT, E., 'American Drone Strike in Libya Kills Top Qaeda Recruiter,' *New York Times*, 28 March 2018, https://www.nytimes.com/2018/03/28/world/africa/us-drone-strike-libya-qaeda.html?searchResultPosition=2

78 SCHMITT, E., 'US Drone Strikes Stymie ISIS in Southern Libya,' *New York Times*, 18 November 2019, https://www.nytimes.com/2019/11/18/world/africa/drone-strikes-isis-libya.html

79 STEWART, P. and LEWIS, A., 'Exclusive: US says drone shot down by Russian air defences near Libyan capital', *Reuters*, 7 December 2019, https://www.reuters.com/article/us-usa-libya-russia-drone-exclusive/exclusive-u-s-says-drone-shot-down-by-russian-air-defences-near-libyan-capital-idUSKBN1YB04W

80 UNITED NATIONS SECURITY COUNCIL, *Letter dated 1 June 2017 from the Panel of Experts on Libya established pursuant to resolution 1973 (2011) addressed to the President of the Security Council*, S/2017/466 (1 June 2017), https://undocs.org/S/2017/466

81 UNITED NATIONS SECURITY COUNCIL, *Letter dated 29 November 2019 from the Panel of Experts on Libya established pursuant to resolution 1973 (2011) addressed to the President of the Security Council*, S/2019/914 (9 December 2019), https://www.securitycouncilreport.org/atf/cf/%7B65BFCF9B-6D27-4E9C-8CD3-CF6E4FF96FF9%7D/S_2019_914.pdf

82 KINGTON, T., 'UAE allegedly using Chinese drones for deadly airstrikes in Libya,' *DefenseNews*, 2 May 2019, https://www.defencenews.com/unmanned/2019/05/02/uae-allegedly-using-chinese-drones-for-deadly-airstrikes-in-libya/

83 HUMAN RIGHTS WATCH, 'Libya: UAE Strike Kills 8 Civilians,' news release, 29 April 2020, https://www.hrw.org/news/2020/04/29/libya-uae-strike-kills-8-civilians

84 BBC STAFF, 'UAE implicated in lethal drone strike in Libya,' *BBC*, 28 August 2020, https://www.bbc.com/news/world-africa-53917791

85 UNITED STATES DEPARTMENT OF STATE, '2020 Country Reports on Human Rights Practices: United Arab Emirates,' accessed at: https://www.state.gov/reports/2020-country-reports-on-human-rights-practices/united-arab-emirates/

86 HARCHAOUI, J., 'Why Turkey Intervened in Libya,' Foreign Policy Research Institute, December 2020, https://www.fpri.org/wp-content/uploads/2020/12/why-turkey-intervened-in-libya.pdf

87 BINNIE, J., 'Turkish UAVs suffer high attrition in Libya,' *Janes*, 12 December 2019, https://www.janes.com/defence-news/news-detail/turkish-uavs-suffer-high-attrition-in-libya

88 Letter dated 29 November 2019 from the Panel of Experts on Libya established pursuant to resolution 1973 (2011) addressed to the President of the Security Council,' https://digitallibrary.un.org/record/3838591?ln=en

89 France 24 Staff, 'France carries out first armed drone strike in Mali', *France24*, 24 December 2019, https://www.france24.com/en/20191224-france-says-it-carried-out-first-armed-drone-strike-in-mali

90 Ministere des Armees, 'Barkhane : 40 000 heures de vol pour le MQ-9 Reaper,' news release, 2 April 2021, https://www.defense.gouv.fr/air/actus-air/barkhane-40-000-heures-de-vol-pour-le-mq-9-reaper

91 Ministere des Armees, 'Commandement de la Défense Aérienne et des Operation Aériennes, Retrospective 2020', https://fr.calameo.com/read/000014334a492f6aba63a

92 Air & Cosmos Staff, 'Défense : la 3D de Barkhane a frappé une centaine de fois en 2020', *Air and Cosmos*, 29 October 2020, https://www.air-cosmos.com/article/dfense-la-3d-de-barkhane-a-frapp-une-centaine-de-fois-en-2020-23798

93 Assemblée Nationale, 'Rapport d'information sur l'Operation Barkhane', No 4089, 14 avril 2021, https://www.assemblee-nationale.fr/dyn/15/rapports/cion_def/l15b4089_rapport-information.pdf

94 Ministere des Armees, 'Point de situation des opérations du 2 au 8 avril', news release, 9 April 2021, https://www.defense.gouv.fr/english/operations/points-de-situation/point-de-situation-des-operations-du-2-au-8-avril/

95 Holland Michel, A., 'Unarmed and Dangerous: The Lethal Applications on Non-Weaponized Drones', *Center for the Study of the Drone*, March 2020, https://dronecenter.bard.edu/files/2020/03/CSD-Unarmed-and-Dangerous-Web.pdf/

96 Akinwotu, E., 'France under pressure to admit responsibility for Mali airstrike', *The Guardian*, 9 April 2021, https://www.theguardian.com/world/2021/apr/09/france-under-pressure-to-admit-responsibility-for-mali-airstrike

97 Reuters Staff, 'U.S. confirms deployment of armed drones in Niger', *Reuters*, 30 July 2018, https://www.reuters.com/article/us-usa-niger-security-drones/u-s-confirms-deployment-of-armed-drones-in-niger-idUSKBN1KK18C/

98 Babb, C., 'US-Constructed Air Base in Niger Begins Operations', *Voice of America*, 1 November 2019, https://p.calameoassets.com/200528160041-e4d0039593cc7c61ae-fa5bc4ca78459c/p46.jpg

99 Altman, H, 'US Africa Command says one of its drones, an apparently armed Gray Eagle, malfunctioned in Niger', *Military Times*, 24 January 2021, https://www.militarytimes.com/news/your-military/2021/01/24/us-africa-command-says-one-of-its-drones-apparently-an-armed-gray-eagle-malfunctioned-in-niger/

100 Ministere des Armees, 'BARKHANE : AFRICOM à la rencontre de la Force Barkhane', news release, 1 March 2021, https://www.defense.gouv.fr/operations/afrique/bande-sahelo-saharienne/operation-barkhane/breves/barkhane-africom-a-la-rencontre-de-la-force-barkhane

101 Gibbons-Neff, T., 'Israeli-made kamikaze drone spotted in Nagorno-Karabakh conflict,' *Washington Post*, 5 April 2016, https://www.washingtonpost.com/news/checkpoint/wp/2016/04/05/israeli-made-kamikaze-drone-spotted-in-nagorno-karabakh-conflict/

102 Fogel, B. and Mathewson, A., 'The next frontier in drone warfare? A Soviet-era crop duster', *Bulletin of Atomic Scientists*, 10 February 2021, https://thebulletin.org/2021/02/the-next-frontier-in-drone-warfare-a-soviet-era-crop-duster/

103 Watling, J. and Kaushal, S., 'The Democratisation of Precision Strike in the Nagorno-Karabakh Conflict,' Royal United Services Institute, 22 October 2020, https://rusi.org/commentary/democratisation-precision-strike-nagorno-karabakh-conflict/

104 Filtenborg, E. and Weichert, S., 'Attack Drones Dominating Tanks as Armenia-Azerbaijan Conflict Showcases the Future of War,' *Daily Beast*, 26 October 2020, https://www.thedailybeast.com/attack-drones-dominating-tanks-as-armenia-azerbaijan-conflict-showcases-the-future-of-war/

105 Gressel, G., 'Military lessons from Nagorno-Karabakh: Reason for Europe to worry,' European Council on Foreign Relations, 24 November 2020, https://ecfr.eu/article/military-lessons-from-nagorno-karabakh-reason-for-europe-to-worry/

106 Welt, C. and Bowen, A. S., 'Azerbaijan and Armenia: The Nagorno-Karabakh Conflict,' Congressional Research Service, 7 January 2021, https://fas.org/sgp/crs/row/R46651.pdf

107 Kofman, M, 'A Look at the Military Lessons of the Nagorno-Karabakh Conflict,' *The Moscow Times*, 21 December 2020, https://www.themoscowtimes.com/2020/12/21/a-look-at-the-military-lessons-of-the-nagorno-karabakh-conflict-a72424

108 Bershidsky, L., 'Drones Have Raised the Odds and Risks of Small Wars,' *Bloomberg*, 30 November 2020, https://www.bloomberg.com/opinion/articles/2020-11-30/drones-have-raised-the-odds-and-risks-of-small-wars

109 Rubin, U. 'The Second Nagorno-Karabakh War: A Milestone in Military Affairs,' The Begin-Sadat Center for Strategic Studies, December 2020, https://besacenter.org/wp-content/uploads/2020/12/184web-no-ital.pdf

110 Günyol, A., '77 Azerbaycan askeri Bayraktar TB2 SİHA Operatörlüğü eğitimini başarıyla tamamladı,' *Anadolu Agency*, 5 February 2021, https://www.aa.com.tr/tr/turkiye/77-azerbaycan-askeri-bayraktar-tb2-siha-operatorlugu-egitimini-basariyla-tamamladi/2135461

111 Hennigan, W. J., 'A fast growing club: Countries that use drones for killing by remote control,' *Los Angeles Times*, 22 February 2016, https://www.latimes.com/world/africa/la-fg-drone-proliferation-2-20160222-story.html

112 BINNIE, J., 'US general dismisses Chinese kit in Africa,' *Janes*, 4 April 2019, https://www.janes.com/defence-news/news-detail/us-general-dismisses-chinese-kit-in-africa

113 NEWS AGENCY OF NIGERIA, 'NAF chief calls for alternative sources of funds for armed forces,' *pulse.ng*, 1 April 2019, https://www.pulse.ng/news/local/naf-chief-calls-for-alternative-sources-of-funds-for-armed-forces/e2l9m5h

114 BINNIE, J., 'Nigeria to ger more armed UAVs from China,' *Janes*, 14 October 2020, https://www.janes.com/defence-news/news-detail/nigeria-to-get-more-armed-uavs-from-china

115 MAZZETTI, M. and Schmitt, E., 'US Expands Its Drone War Into Somalia,' *New York Times*, 1 July 2011, https://www.nytimes.com/2011/07/02/world/africa/02somalia.html?pagewanted=1&_r=1&ref=todayspaper

116 TBIJ STAFF, 'Drone Warfare', *The Bureau of Investigative Journalism*, accessed on 1 September 2021, https://www.thebureauinvestigates.com/projects/drone-war

117 SAVAGE, C., COOPER, H. and SCHMITT, E., 'US Strikes Shabab, Likely a First Since Trump Relaxed Rules for Somalia,' *New York Times*, 11 June 2017, https://www.nytimes.com/2017/06/11/us/politics/us-airstrike-somalia-trump.html

118 SNOW, S., 'US Strikes ISIS fighters in Somalia,' *Air Force Times*, 25 October 2019, https://www.airforcetimes.com/flashpoints/2019/10/25/us-strikes-isis-fighters-in-somalia/

119 SZAKOLA, A., 'Iran admits conducting drone strikes in Syria,' *Business Insider*, 26 September 2016, https://www.businessinsider.com/iran-drone-strikes-syria-2016-9

120 Ibid

121 GORDON, M., 'American Warplane Shoots Down Iranian-Made Drone Over Syria,' *New York Times*, 20 June 2017, https://www.nytimes.com/2017/06/20/world/middleeast/american-warplane-shoots-down-iranian-made-drone-over-syria.html

122 KERSHNER, I., 'Iranian Drone Launched From Syria was Armed, Israel Says,' *New York Times*, 14 April 2018, https://www.nytimes.com/2018/04/14/world/middleeast/syria-iran-israel-drone.html

123 TASS STAFF, 'Russia's Orion attack drone arrives for troops after Syria experience – source,' *TASS*, 1 November 2019, https://tass.com/defence/1086508

124 MLADENOV, A., 'Russia still has sights set on Orion,' *Shephard News*, 23 March 2021, https://www.shephardmedia.com/news/uv-online/premium-russia-still-has-sights-set-orion/

125 RIA NOVOSTI STAFF, 'В Сирии испытали уникальный метод наведения ударных беспилотников,' *RIA Novosti*, 4 April 2021, https://ria.ru/20210404/siriya-1604135512.html

126 YILDIRIM, G, 'SİHA'ların sayısı da silahı da arttı', *Anadolu Agency*, 13 October 2017, https://www.aa.com.tr/tr/gunun-basliklari/sihalarin-sayisi-da-silahi-da-artti/934484

127 PRESIDENCY OF DEFENCE INDUSTRIES, 'MİLLİ SİHA BAYRAKTAR, KATAR GÖREVİNE HAZIR', news release, 1 February 2019 https://www.ssb.gov.tr/website/ContentList.aspx?PageID=1662

128 ÖZÇELIK, N, 'Askeri harekâtların kuvvet çarpanı: İnsansız hava araçları', *Anadolu Agency*, 19 April 2018, https://www.aa.com.tr/tr/analiz-haber/askeri-harek%C3%A2tlarin-kuvvet-carpani-insansiz-hava-araclari/1122660

129 DAILY SABAH STAFF, 'Drones turned fortune of Afrin operation, prime minister says,' *Daily Sabah*, 23 March 2018, https://www.dailysabah.com/defence/2018/03/23/drones-turned-fortune-of-afrin-operation-prime-minister-says

130 KASAPOGLU, C. and ULGEN, S., 'Operation Olive Branch: A Political Military Assessment', Center for Economics and Foreign Policy Studies, 30 January 2018 https://edam.org.tr/en/operation-olive-branch-a-political-military-assessment/

131 GALL, C., 'Turkey Declares Major Offensive Against Syrian Government,' *New York Times*, 25 April 2021, https://www.nytimes.com/2020/03/01/world/middleeast/turkey-syria-assault.html

132 ROBLIN, S., 'Turkish Drones and Artillery Are Devastating Assad's Forces in Idlib Province– Here's Why,' *Forbes*, 2 March 2020, https://www.forbes.com/sites/sebastienroblin/2020/03/02/idlib-onslaught-turkish-drones-artillery-and-f-16s-just-destroyed-over-100-armored-vehicles-in-syria-and-downed-two-jets/?sh=4fa8e6156cd3

133 KASAPOGLU, C., 'Turkey's Drone Blitz Over Idlib,' *The Jamestown Foundation*, 17 April 2020, https://jamestown.org/programme/turkeys-drone-blitz-over-idlib/

134 ALSAAFIN, L., 'Turkey shoots down two Syrian fighter jets over Idlib,' *Al Jazeera*, 1 March 2020, https://www.aljazeera.com/news/2020/3/1/turkey-shoots-down-two-syrian-fighter-jets-over-idlib

135 TURKISH MINISTRY OF NATIONAL DEFENCE (@tcsavunma), 'Asil milletimiz emin olsun ki; İdlib'de barış ve huzur için toprağa düşen Aziz şehitlerimizin hesabını sormaya devam edeceğiz!,' Twitter, 29 February 2020, https://twitter.com/tcsavunma/status/1233855282486370311

136 GALL, C., 'Turkey Declares Major Offensive Against Syrian Government,' *New York Times*, 1 March 2020, https://www.nytimes.com/2020/03/01/world/middleeast/turkey-syria-assault.html

137 CNN TURK STAFF, 'İşte yerli İHA ve SİHA'ların üssü! CNN TÜRK ekibi savunmanın kalbinde,' *CNN Turk*, 3 March 2020, https://www.cnnturk.com/video/turkiye/iste-yerli-iha-ve-sihalarin-ussu-cnn-turk-ekibi-savunmanin-kalbinde

138 OZKOK, E., 'Şu ıstırap karesini kaç kişi kimden ve ne zaman öğrendi', *Hurriyet*, 3 March 2020, https://www.hurriyet.com.tr/yazarlar/ertugrul-ozkok/su-istirap-karesini-kac-kisi-kimden-ve-ne-zaman-ogrendi-41459664

139 Atlas, M., 'Erdogan obišao bazu turskih bespilotnih letjelica,' *Anadolu Agency*, 3 February 2018,
https://www.aa.com.tr/ba/turska/erdogan-obi%C5%A1ao-bazu-turskih-bespilotnih-letjelica/1053619

140 Mason, R., 'UK to fly military drones over Syria,' *The Guardian*, 21 October 2014, https://www.theguardian.com/uk-news/2014/oct/21/uk-to-fly-military-drones-over-syria

141 MacAskill, E., 'Briton killed in drone strike on Isis 'posed serious threat to UK',' *The Guardian*, 26 April 2017,
https://www.theguardian.com/uk-news/2017/apr/26/briton-killed-in-drone-strike-on-isis-posed-serious-threat-to-uk-reyaad-khan

142 Drone Wars staff, 'UK Drone Strike Stats', Drone Wars UK, accessed on 1 September 2021, https://dronewars.net/uk-drone-strike-list-2/

143 Yeni Safak Staff, 'Yerli İHA ilk harekatında 5 teröristi vurdu,' *Yeni Safak*, 8 September 2016,
https://www.yenisafak.com/gundem/yerli-iha-ilk-harekatinda-5-teroristi-vurdu-2528552

144 Risen, J. and Miller, J., 'CIA Is Reported to Kill a Leader of Qaeda in Yemen,' *New York Times*, 5 November 2002,
https://www.nytimes.com/2002/11/05/world/threats-responses-hunt-for-suspects-cia-reported-kill-leader-qaeda-yemen.html

145 Mazzetti, M., 'US is Intensifying a Secret Campaign of Yemen Airstrikes,' *New York Times*, 8 June 2011,
https://www.nytimes.com/2011/06/09/world/middleeast/09intel.html?_r=3&partner=rss&emc=rss

146 Mazzetti, M., Schmitt, E. and Worth, R. F., 'Two-Year Manhunt Led to Killing of Awlaki in Yemen,' *New York Times*, 30 September 2011,
https://www.nytimes.com/2011/10/01/world/middleeast/anwar-al-awlaki-is-killed-in-yemen.html

147 Mazzetti, M. and Callimachi, R., 'Top Qaeda Figure Dies in Yemen Drone Strike,' *New York Times*, 7 May 2015,
https://www.nytimes.com/2015/05/08/world/middleeast/top-qaeda-figure-dies-in-yemen-drone-strike.html?searchResultPosition=7

148 Fielding-Smith, A. and Purkis, J., 'Drone Strikes in Yemen,' Bureau of Investigative Journalism, last updated February 2020, accessed on 5 April 2021, https://www.thebureauinvestigates.com/projects/drone-war/yemen

149 DeYoung, K. and Miller, G., 'White House releases its count of civilian deaths in counterterrorism operations under Obama,' *Washington Post*, 1 July 2016,
https://www.washingtonpost.com/world/national-security/white-house-releases-its-count-of-civilian-deaths-in-counterterrorism-operations-under-obama/2016/07/01/3196aa1e-3fa2-11e6-80bc-d06711fd2125_story.html

150 Lawfare staff, 'Legality of Targeted Killing Programme under International Law,' Lawfare (blog), accessed on 5 April 2021,
https://www.lawfareblog.com/legality-targeted-killing-programme-under-international-law

151 Michel, A. H., 'Drone Litigation Updates,' *Center for the Study of the Drone*, 25 March 2019,
https://dronecenter.bard.edu/drone-litigation-updates/

152 Shaif, R., and Watling, J., 'How the UAE's Chinese-Made Drone Is Changing the War in Yemen,' *Foreign Policy*, 27 April 2018,
https://foreignpolicy.com/2018/04/27/drone-wars-how-the-uaes-chinese-made-drone-is-changing-the-war-in-yemen/

153 Associated Press staff, 'UAE Dismantles Eritrea Base as it Pulls Back After Yemen War,' *Associated Press*, 18 February 2021,
https://www.voanews.com/africa/uae-dismantles-eritrea-base-it-pulls-back-after-yemen-war

154 Hubbard, B. and Isabel Kershner, I., 'Sinai Blasts Kill Up to 5 Thought to Be Militants,' *New York Times*, 9 August 2013,
https://www.nytimes.com/2013/08/10/world/middleeast/sinai-blasts-kill-up-to-5-islamic-militants.html

155 Wainer, D., Ferziger, J. and Feteha, A., 'Old Mideast Foes Unite Over Gas Deals and Fighting Militants,' *Bloomberg*, 11 July 2016,
https://www.bloomberg.com/news/articles/2016-07-10/old-middle-east-foes-unite-over-gas-deals-and-fighting-militants

156 Kirkpatrick, D. D., 'Secret Alliance: Israel Carries Out Airstrikes in Egypt, With Cairo's OK,' *New York Times*, 3 February 2018,
https://www.nytimes.com/2018/02/03/world/middleeast/israel-airstrikes-sinai-egypt.html

Index